A LIFE IN NATURE

A LIFE IN NATURE

Sir Peter Scott

SPHERE

First published at *Happy the Man: Episodes in an Exciting Life*
in Great Britain in 1967 by Sphere Books
This edition published in 2021 by Sphere

1 3 5 7 9 10 8 6 4 2

A CIP catalogue record for this book
is available from the British Library.

ISBN 978-0-7515-8335-9

Typeset in Bembo by M Rules
Printed and bound in Great Britain by Clays Ltd, Elcograf S.p.A.

Papers used by Sphere are from well-managed forests
and other responsible sources.

Sphere
An imprint of
Little, Brown Book Group
Carmelite House
50 Victoria Embankment
London EC4Y 0DZ

An Hachette UK Company
www.hachette.co.uk

www.littlebrown.co.uk

Contents

Introduction ix
Timeline xi

PART I NATURALIST

1 Brown and Green 5
2 A Rare Meal 9
3 The Ethics of Shooting 15
4 Badgers and the Animals of the Wiltshire Woods 18
5 The Introduction of Exotic Species 25
6 Food Production and Conservation 31
7 Search for Salvadori's Duck 35
8 Skin-Diving: A New World 49
9 Sightseeing in East Africa's Game Parks 59
10 Conservation and Africa 80
11 To Record the Moment – or Enjoy It? 87
12 Anabel 88
13 The Mysterious Sense of Direction 93
14 The Aura 107
15 European Travels 110
16 Conscience 145
17 The Guides Return 146
18 The Flowers of the Eagle's Fell 156
19 Slimbridge Discovered 175
20 Bewick's Swans at Slimbridge 180
21 How Much Does It Matter? 188

22 A Voice Crying in the Wilderness? 192
23 The Best the Arctic Can Offer 195
24 Wild Music 206
25 South America – 1953 208
26 Saving a Goose from Extinction 223
27 Galapagos Expedition 227
28 September in the Country 234

PART II NAVAL OFFICER

29 Nature of Fear 243
30 Camouflage 263
31 Fire at Sea 268
32 A Battle – and a Collision 287
33 Dawn off Cherbourg 299
34 The Story of SGB9 318

PART III SPORTSMAN

35 'Gold C' Completed 329
36 Straight and Level, Please 338
37 To Make the Spirit Soar 342
38 Prince of Wales's Cup and Olympic Games 344
39 The Trapeze 362

PART IV HAPPY THE MAN

40 The America's Cup 371
41 The Spirit of Adventure 375
42 Mr Spriggs and the Crane 379
43 Britain's Canals 389
44 Reflections 392
45 The Boy David 396
46 World Champion Skater? 401
47 Venice, 1939 403

48 Journey to the South Pole 405
49 Happy the Man 416

Dates and Sources of Original Publication 427

Introduction

Introduction

I had the extremely memorable privilege of meeting Peter Scott in 1988. He had been a major influence in my early days with his natural history broadcasts on radio and television, as I know he had been with countless others of my generation. My parents took me to Slimbridge when I was nine and I recall to this day the joy I experienced feeding the amazing collection of wildfowl. Little did I know then that I would one day have another great privilege, of leading the organisation so dear to his heart, the Wildfowl & Wetlands Trust, and having as my office that famous studio room with its window looking over the Rushy Pen and the wintering swans, geese and ducks.

The world is so frequently driven forward by the vision of an individual. In my view, and it is one shared by many others, Peter Scott was one of the most influential conservationists of the last century. He was a man of great vision and he has left a lasting legacy which now is even more relevant to the world we need to cherish so much.

Peter was one of those people with the combination of talent, drive and determination to achieve, and did so with whatever he undertook. He was an accomplished artist, an Olympic yachtsman and an ice skater, a gifted communicator, one of the first natural history broadcasters, and a champion glider pilot among so many other pursuits and attributes. But it is his passion for the natural world and his vision for its conservation that is, for me, his greatest legacy.

Peter Scott was among a group of far-sighted individuals who saw what was happening to the natural world and were determined to do something about it. But he was also far ahead of his time in recognising, more than seven decades ago, a fact that is now, in this world increasingly disconnected from nature, so vital to our future on this wonderful planet. To quote him, 'The most effective way to save the threatened and decimated natural world is to cause people to fall in love with it again with its beauty and its reality.' Surely, Peter was the architect of modern conservation. He reflected then on our great challenge now in our digitally dominated and environmentally challenged world.

This book is a selection of Peter's writings and broadcasts from the first half of his life, when the seeds of his passion for conserving the natural world were being planted. It covers his wartime experiences in the Navy, early travels, and his time as a wildfowler. But it is his achievements for conservation that for me stand out. Having established the Severn Wildfowl Trust, now the Wildfowl & Wetlands Trust, on the banks of the River Severn in 1946, he went on to be an active vice-president in the newly formed International Union for Nature Conservation in 1948, and then to be a founder of WWF, the World Wide Fund for Nature, in 1961. A man who thought globally, he was a major player in the formation in 1971 of the first global convention for conservation, the Ramsar Convention on Wetlands. Now with 170 countries' governments as signatories, it is the key driver to the protection of the world's wetlands, so important for biodiversity and climate management.

Peter Scott saw the need for, and value of, education and he was at the forefront of environmental education in the UK, especially focusing on the younger generations to take change forward. This continues at the wetland centres he established across the country. His last great vision, to take nature to the people, resulted in the internationally acclaimed London Wetland Centre, where connection with nature and education are the primary aims.

Personally, I prefer not to think of his legacy as though it is something of the past. It is about the present and the future. We now need Peter Scott's same degree of foresight, vision, drive and determination more than ever.

Martin Spray, CBE, DSc
March 2021

Timeline

SIR PETER MARKHAM SCOTT, CH, CBE, DSC, FRS

1909 Born London, 14 September.
1912 Father (Captain Robert Falcon Scott) died on return from South Pole.
1919 Went to West Downs School.
1923 Went to Oundle School.
1927 Went to Trinity College, Cambridge.
1929 First drawings published (*Country Life*).
1931 Went to State Academy School, Munich.
1932 Went to Royal Academy Schools, London.
1933 First exhibition of paintings (Ackermann's Galleries, London).
1934 Established collection of tame waterfowl at East Lighthouse, Sutton Bridge, Lincolnshire.
1935 First book published (*Morning Flight*).
1936 Won bronze medal at Olympic Games (single-handed sailing); led British dinghy team to Canada.
1937 Won Prince of Wales's Cup (for 14′ dinghies); went on expedition to Caspian Sea area.
1938 Won Prince of Wales's Cup; went on expedition to Lapland.
1939 Volunteered for Royal Navy.
1940 Served in HMS *Broke* in Battle of France, and in Battle of Atlantic (until 1942).
1941 Mentioned in despatches.

1942 Married Elizabeth Jane Howard (marriage later dissolved); awarded MBE; joined light coastal forces in command Steam Gunboat HMS *Grey Goose*; took part in Dieppe Raid; mentioned in despatches.
1943 Daughter Nicola born; appointed to command Steam Gunboat Flotilla; awarded DSC and Bar and mentioned in despatches.
1944 Joined staff of C in C Portsmouth; based in Normandy Bridgehead after D-Day; commanded force of American ships.
1945 Appointed to command frigate HMS *Cardigan Bay*; met surrendering German E-boats off Thames Estuary; fought General Election as Conservative candidate for Wembley North (lost by 435 votes).
1946 Founded the Severn Wildfowl Trust (later the Wildfowl Trust); President, Home Counties North Young Conservatives; first of regular broadcasts in *Nature Parliament*; won Prince of Wales's Cup for third time.
1948 Chairman, Olympic Yachting Committee.
1949 Explored Perry River region of Arctic Canada.
1950 Expedition to Lapland.
1951 Married Philippa Talbot-Ponsonby; expedition to central highlands of Iceland.
1952 Daughter Dafila born.
1953 Awarded CBE as Director of the Wildfowl Trust; expedition to central highlands of Iceland; first edition of television programme *Look*; travelled to South America.
1954 Son Falcon born; winner of sailing speed record, Cowes.
1955 Became President of International Yacht Racing Union.
1956 Travelled to Uganda; started gliding; Chairman, Olympic Games International Jury for Yachting (and in 1960, 1964).

1957 Travelled to Australia, New Guinea, New Zealand, Fiji, and Hawaii; won gliding Silver C Badge; Vice-President, International Union for Conservation of Nature.
1958 Gliding 'Gold C' International Badge.
1959 Travelled to Trinidad, Panama, Ecuador, Galapagos Islands.
1961 Travelled to Kenya and Tanganyika; closely involved with establishment of the World Wildlife Fund.
1962 Gliding Third Diamond Badge; 100th edition of *Look*, with HRH The Duke of Edinburgh as guest.
1963 British National Gliding Champion; Chairman, Survival Service Commission of International Union for Conservation of Nature.
1964 Travelled to United States and Japan; unsuccessfully challenged for the America's Cup at the helm of *Sovereign*.
1965 Travelled to the Bahamas and India and Thailand.
1966 Travelled to Antarctic; special edition of *Look* for National Nature Week with the Duke of Edinburgh.
1973 Knighted for services to the conservation of wild animals.
1973 Appointed Chancellor of the University of Birmingham.
1979 Awarded honorary doctorate University of Bath.
1987 Appointed Companion of Honour; elected Fellow of the Royal Society.
1989 Died Bristol, 29 August.
1996 Scott's life and work in wildlife conservation was celebrated in a major BBC Natural World documentary, *In the Eye of the Wind*, narrated by Sir David Attenborough.
2004 Profiled in the BBC Two series *The Way We Went Wild*.
2006 BBC Four documentary, *Peter Scott – A Passion for Nature*.

A LIFE IN NATURE

PART I

NATURALIST

1

Brown and Green

Three hundred years ago the salt-marshes of the Wash passed slowly into the great fresh-water marshes of Cambridgeshire.

In those days the grey mud stretched between the channels and sent arms up each side of them many miles inland. And, when the tide flowed in, the fresh water was piled up and spread through the reed-covered swamps to the little pools where the wild duck rested during the day. When the rain came in winter the flood-water shimmered over the fens as it covered the tops of the grass stems, and men lived at Ely and called it an island.

But now the sea is allowed a paltry mile or two of mud and salt-marsh to play with, and then it is shut out by long lines of green bank which protect the brown peaty fields of potatoes.

I went there once in June to see again the place where I had spent so many happy days in winter. As I sat on the bank in the thick, luscious, green grass and looked across the fields to the farm, I was filled with wonder at a scene so green. The green rows of potato plants throwing the field into weird patterns of perspective, the few corn-fields still green like the grass of the sea-wall on which I sat, and the stunted elm trees thick in foliage, amongst which the little farm took shelter from the winter winds; it was all green and rich.

And as I looked my mind went back to a day in January when I had sat with a good friend in just the same place and we had been eating our breakfast. We had waited at dawn on the salt-marsh for the wild geese to flight, and as the light had come the grey rainy skies had lifted to the west so that the atmosphere was wonderfully clear.

From the sea-wall the brown potato fields stretched for miles, broken here and there by patches of lighter stubble. A plough at work in the distance looked clear-cut, and the little farm amongst the trees seemed so close that it was hard to believe that two forty-acre fields lay between. It was all brown and purple. And far away was a black speck in the sky, which seemed to dissolve and then form again. It was the thickest part of a pack of pink-footed geese. They had gone inland at dawn, and they were circling round and round over their feeding-grounds. Presently some more stragglers flew in from the mud and went on to join them.

Whilst we watched a grey cloud came past in front of us, and below it trailed a curtain of rain – dark streaks reaching down to the ground. It blotted out the farm and the trees and blended with the dark-brown earth so that all the edges and hard lines were softened and blurred.

The curtain swept by, and the land was clear again; so clear that the little farm looked closer than ever. Away in the distance still circled the geese. Somewhere inland a yellow shaft lit up the fens where the sun was shining through the clouds. It seemed that all the light was reflected along from this opening under the grey roof of rain-clouds. The rain-drops glistened on the grass stems, each with a spot of deep brown, the brown of the fields.

The colour was brown, and as we stood up on the sea-wall we looked back at the salting behind us. That was olive-green, but beyond it the mud was brown and the green of the salt-marsh plants seemed to blend with the brown. The succulent plants had a purplish sheen on them and the tips of the grass stems were dead and dry and brown.

But in June the grass was long and thick on the saltings, just as it was on the bank, and out on the edge of the tide there were godwits and redshanks calling their summer trill – and the colour was green.

Near where I sat was a row of straggly willow trees, and as I looked at them I thought of how many things I had enjoyed within sight of them. 'Why, trees,' I said, 'you remember that night under the moon when the geese came so low and the foggy morning and the hard frost.' The silver-green leaves whispered their answer in the summer breeze.

In winter, when we tire of the great, flat, unbroken sweep of brown, and we need some variation to relieve the monotony, then we go to the coast of Norfolk. Above the marshes rise wooded hills with birch and pine and rhododendrons – for the country is stocked with pheasants. Down below is the green fresh marsh, green all through the winter. There is a shingle bank which shines orange in the sun, and it is crowned with marram grass. Out on the mud there is green sea-grass.

There are little bathing huts there too. They look forlorn and absurd when the north-east wind blows and the wild geese beat in against it low over their little corrugated roofs. The end one is white and has a red roof. It is the best cover on the beach.

The marsh behind is the haunt of white-fronted geese.

Perhaps it is fitting that this yellow shingle and green fresh marsh should be their home. Of all our geese they are the most brilliant and colourful. White forehead, black bars on their breasts, and orange legs. But with all that, I prefer the grey pinkfeet. For them there are the browns of the potato land.

That land must always have been brown even in the days when the tide came in – for a great rippling sea of reeds must have carpeted it with here and there an open pool, where mallards and teal lived.

But in those days there could have been no pinkfeet there: for they love the hard black earth and the mushy potatoes to nibble. Now they come in their thousands from October to March. And every morning the great skeins come in, high in the air, and circle round over their chosen fields, and every evening with great clamouring they fly out through the frosty blackness as they come in to feed – until the brown land is full of geese.

But behind the green fresh marsh of the whitefronts the land rises to heather-covered slopes and then the woods begin.

I went there, too, in June. In a little hollow of heather I found a pool of fresh water, and by the pool were two rhododendron bushes in flower, and all around sat shelducks; some of them got up and circled round the little hollow and over the place where I hid in some bracken. As they flew they laughed at me with their deep mock-human call. Here, where the sand is yellow amongst the purple of the heather, the shelducks find rabbit holes to nest in. Outside a hole near me were web-footed tracks in the soft sand. It was hot and the heather smelled good.

From the hill I could look out over the marsh to the mud and the sea and the little bathing huts and the great superstructures of the shingle works. With the sight of them came memories crowding and hustling, so that one had scarcely been enjoyed again – and mourned – when the next came pushing on, until I could no longer check each one. A great flood of them came upon me, half joy, half sadness – joy that I had spent those times and that I could almost live them through again, and sadness that they were passed, those individual days when our disappointments and mistakes formed part of our treasured memory, and those with whom that happiness had been shared could not perhaps be there again to share it.

From *Morning Flight*

2

A Rare Meal

Tuesday June 28. At about six o'clock in the morning it began to rain. It was only a light spotting, at first; then a drizzle. The Eskimos seemed to be sleeping through it, but at length it became too brisk for them. I beckoned them into the tent, and in they came. So, now, as I write, there are four of us in the tent, which was made for two. The rain is heavy and continuous, and the tent leaks considerably. We have not yet reached the Ross's Goose Lake – and then there is the return journey and we have already used half our food. We wonder how long our food and the rain will last. Already we have lain here all morning. It is now one o'clock, and the Eskimos are still snoring comfortably. Harold is asleep, and I am writing – and we still haven't had breakfast. Jimmy has a pretty bad cough, which sounds fairly tubercular. Both of them sleep soundly, snoring only slightly. They sleep curled up, and the movements of their sleeping companions seem to disturb them not at all.

We lay in the tent, having a meal of biscuits and candy and jam, without heating anything until about 4 p.m. when it stopped raining. So we emerged to find a slightly more promising sky, with high strato-cumulus clouds. But it was only temporary. We cooked a cup of cocoa, and then set off south-westward again to look for the Ross's geese.

Almost at once it began to rain – a soft, persistent, utterly wetting rain. It had become important that we should kill some game for food, as the situation was already bad. Apart from some dried potatoes, some macaroni, and a couple of cans of fruit, we were more or less out. So it was rather disastrous when I missed some lesser snow geese which passed within range.

After an hour or so we climbed a little hill. From the top we saw a marsh with a large number of Ross's geese – at least fifty – feeding on it. There was also a herring gull on a nest, which we did not visit as it lay out of our way. Patsy went on ahead like a mountain goat, and up the next precipitous little peak. We climbed behind him up the rocky hillock and there, beyond, was a much bigger marsh, and beyond it a much bigger lake than any of the ones we had passed. There were several bunches of white geese. We sat down on top of the hill to look them over with the glasses through the rain. They were nearly all Ross's geese, but there were a few lessers mixed in with them. We looked beyond the marsh towards the lake – and, there it was, the Ross's goose breeding colony. In the lake were five little islands, and all over these were the white dots of the sitting geese with their ganders. They were quite thick on the islands – hundreds of them.

The islands were less than a hundred yards long, and we made counts of the geese. We could only see one side of each island, but we estimated what was hidden from what we could see. Over to the left was a sixth island, which seemed to be in another lake, though we afterwards discovered that the two lakes were connected by a narrow strait. On the six islands we estimated that there were perhaps 350 pairs of birds. The nests were thickest on the smallest island, which was quite a low mound. The other islands were rocky mounds rising ten or twenty feet out of the lake.

We walked down across the marsh to the nearest point of shore opposite the nearest island, and sat down in the shelter of a big square rock. From here I took some film with the telephoto lens, and then sat and watched the birds through the binoculars.

There were probably fifty Ross's nests in front of us, at a range of 200 yards. The intervening lake appeared to be ice-free, although there may have been submerged ice; and indeed our theory to account for most of these lakes being open was that, in this more hilly region, there was a faster run-off of melting snows which flooded down on to the ice of the lakes, covering it with several feet of water.*

In some parts of the island the nests seemed to be no more than five or six feet apart. There were two lesser snows and two lesser Canadas sitting on nests among the Ross's; and there was a single blue goose – apparently immature – among the noisy bunch of seemingly non-breeding lesser snows, standing at one end of the island. There was also an attendant pair of glaucous gulls. A good deal of squabbling seemed to be going on between ganders.

It was a fascinating sight, and we sat there in the rain until it began to get cold. It was sad that the light was so poor for photography, and also extremely frustrating that we could not get across to the islands. The Eskimos indicated that the ice would not be more than three feet down, and had the light been good for photography I think I would have tried to wade over as long as it was not above the waist. But under the circumstances we decided to leave it alone. We trust we shall get to the lake again with Jim Bell and the canoe, or else to another of the colonies. This is the colony of which Gavin had been told by the natives, but we were the first white men ever to see it. We have decided to call it Lake Arlone after Harold's wife.

We walked over to a knoll, which gave a view of the single island in the other half of the lake. On the way we flushed a Buffon's skua from a lemming which it had lately killed. Among about fifty Ross's nests on that island were two lesser Canadas

---

*We afterwards found that this was wrong. The speed of opening up of lakes appears to be directly related to their depth, and these lakes were all absurdly shallow – silted up to a level bottom about three feet down, if as much.

sitting. There was also a blue goose, but it may have been the same as the one we saw earlier.

We had the feeling that we were seeing a very beautiful place at its worst, for the raw chill of the wind and the dark sky and heavy rain cast a gloom over it all. We thought out where, and how, we would drive the geese for banding them later on. There was a small lake – scarcely more than a flash – in the marsh, and we planned to use that in some way, driving the birds first to it, and then setting the net between them and the big lake. There were some pintails on the flash, and one or two long-tailed ducks. There was also a pair of king eiders on one of the smaller islands in the big lake. Two of such islands were rather unexpectedly quite free of goose nests.

As we walked back across the marsh towards our camp, the food situation had become desperate. After two inexcusable misses, I had only two cartridges left (No. 7 shot).

We saw a cock ptarmigan, and Jimmy shot it with the .22 rifle. Then we saw a small bunch of geese at the head of the marsh. We divided the party, and I went to head off their retreat to the lake while the rest went on. The plan worked perfectly, and the geese came down-wind straight over my head. The first fell beside me, and the second carried on for 200 yards and fell dead. The first was a lesser snow, the second a Ross's goose. Quite genuinely, and without stretch of imagination, we had been forced to shoot a Ross's goose for food. Both geese were immature females, which is interesting in showing that the non-breeding birds hang about in the immediate vicinity of the breeding ground. No doubt the flock of fifty which we had seen earlier were non-breeders. Harold is very doubtful whether the three-year-old breeding rule holds good in the wild state, and believes that two-year-olds commonly breed. This is a point which should be easy enough to elucidate by means of ringing. It could probably be done most easily with Canada geese, especially in the colonial nesting sites in Idaho and Utah, using coloured rings.

Harold did not see the second goose fall, and the rest of the party walked on, while I went back 200 yards towards the lake to pick up the Ross's goose, which I found without difficulty. When I turned towards camp again, I could see no sign of Harold or the Eskimos. Actually, they had walked to the nearest bluff at the edge of the marsh and were sitting waiting on the rocks, where they were hard to see. I thought they had gone straight on over the hill and out of sight, and for a while I harboured hard thoughts of their ungratefulness for delivery from starvation. However, they were suitably elated when I came up with them, and I heard from Harold that Patsy had said to him 'Thank you!' in English when the first goose dropped.

We wandered back to camp in a sodden condition, and Harold hastily skinned the two geese with my hunting-knife – which had been used for opening tins. He cut off the breasts of the birds and cooked them up with dried potatoes. Then we climbed into the low tent (it is a khaki tent, three and a half feet high at its highest point), got into our sodden sleeping-bags and called in the Eskimos. Then we had a shock. Patsy came in and said that Jimmy had gone back to the canoe. This left horrible visions of Harold and me struggling under added loads all the way home. On the other hand, it seemed that he might have feared for the safety of the canoe and an increased flood as a result of the rain, for we had not pulled it up to an entirely safe height above the river, and there was no doubt that the Eskimos had not expected to be away for so long. We had a hasty discussion, and thought perhaps that we should break camp at once and start for home, so as to have the assistance of Jimmy. Anyway, Jimmy should be recalled. Patsy understood, and went out; then came back to say that Jimmy was coming. Jimmy took some time returning, and arrived with a ptarmigan cock. When he got back, it seemed that they had both mistaken the long time we had taken getting our wet clothes off in the tent for an intention to leave them out in the rain all night, and rather than that Jimmy had set off to sleep under the canoe – eleven miles away! As soon as we made it clear we intended him to come into the tent, he crept in behind Patsy and we all went to sleep.

We had planned to sleep for four hours, and start back at 4 a.m. Well before that I found the space so cramped that I could not get any comfortable rest. The wet caribou-skin garments and the seal-skin boots smelt much stronger than before. The inside of the tent was running with rain and condensation, and it was quite impossible with the four of us to avoid touching the top. My outer sleeping-bag was getting pretty sodden; so, just before four o'clock on finding Harold awake, too, and the rain stopped, we got up, cooked some macaroni, and set off at 5 a.m. on the long trek home.

From *Wild Geese and Eskimos*

3

The Ethics of Shooting

During a goose-shoot the six or seven 'guns' were standing in a cluster when a single goose flew over. Each man (and I was one) raised his gun and fired two barrels. Twelve shots went off at the goose, which staggered in the air, flew on, and then began to lose height. It came down far out on the mud flats in a place quite inaccessible because of the soft mud which in parts of that estuary amounts almost to quicksand. As it landed, I watched with my glasses, and saw that both its legs were broken. It crash landed, came to rest quivering on its belly, and put its head up. There was nothing that any of us could do. It was 500 yards away and inaccessible. We repaired later to the hotel for lunch, and a very good lunch it was; I remember that the cherry brandy was excellent. In the afternoon I recall that our marksmanship was a little below the standard of the morning. As we passed along the sea-wall I saw with binoculars that the bird still sat out on the mud, its neck still raised. At dusk we went back to tea, a bath before an excellent dinner, and after it a comfortable bed.

On the following day I came out again to watch the geese. The shooting of course was over, and many of the geese were feeding in the fields as if nothing had happened. But out on the mud, in precisely the same place (for there were small neap tides at the time which did not cover the high sand), still sat the goose with the broken legs. 'What right,' I said to myself, 'have we men to do this to a bird for our fun – to impose that kind of suffering? I should not want this for a sworn enemy, and that goose was not my enemy when I shot at him – although I was his.'

'That kind of suffering,' I had said – but what kind of suffering was it, for without being a bird how could one know what pain they felt? Perhaps this is a question which will never be adequately answered, but there are, of course, observations which bear upon it. We know that wounds which would incapacitate a man do not do so to a bird. We know that cockerels will feed unconcernedly immediately after a major abdominal operation. We know that although forty-one per cent of a sample of 825 adult pink-footed geese were shown by X-rays to be carrying shot somewhere in their bodies, the average weight of the forty-one per cent was no lower than the weight of the other fifty-nine per cent, and weight is one of the most sensitive barometers of a bird's well-being. It has been held, and I believe rightly, that pain in birds is something quite different from, and much less than, pain as we know it in man. I mean this in the physiological sense. But psychologically too, it is clear that animals suffer less, in that they are unable to reason about it and to anticipate pain or death. They do not think, 'Perhaps this wound is mortal, perhaps I shall die.' But in spite of all this the goose with the broken legs was upsetting.

I took the argument a stage further. I had in the past derived enormous enjoyment, good health, interest, and aesthetic pleasure from being out on the marshes at dawn and dusk and under the moon, in rain and gale and frost. The birds with their beauty and wildness had been an endless source of delight. The difficulty of outwitting them, the discomforts, the occasional dangers – these, and not the killing, made the sport of wildfowling one of the most exciting in the world. I was not interested in 'the ultimate sanction'. I would have given anything (and I wonder how many others feel the same) for a handy portable device which for the same output of knowledge, skill and manual dexterity would bring a bird to hand alive and unharmed. It was inescapable to me that, had I never been interested in wildfowling, my enjoyment, and also my knowledge of the birds, would have been much less. Was there not a balance here? Perhaps so long as a man is deriving so much good as to offset the bad, the balance might still fall in favour of the wildfowler. If I were advising a young boy would I not say to him: 'Of course you will enjoy wildfowling; it will bring you unparalleled thrills; and if you never experience those thrills you will probably never enjoy the birds themselves to quite the same extent, for you will not learn the subtle goose/rook, wigeon/starling distinction, the subtle difference between man's attitude to his traditional quarry and his attitude to all other birds'?

I am doubtful whether this argument that I might offer is sound, but I am quite sure that as soon as the doubts and the disquiet prevent you from enjoying shooting there can no longer be any good reason for going on doing it. So I have sold my guns, and I no longer shoot. It is easy for the young and keen to say, 'Well, that's all right for him, he's shot all the geese he wants and now he says he doesn't think it's a good thing.' But that merely confuses the issue; a dog-in-the-manger is something different again, something which I hope never to be. It can be argued that the final decisions should be made in the conscience of the individual. 'Do I think this right, do I think this wrong? How much does my decision affect other people? Do I enjoy it so much that the overall good offsets the bad? Or would it be better to give it up?' Having decided to give it up the situation is not quite simple. If cruelty exists on a level comparable, say, with cruelty to children, then more is demanded than personal abstention. But if we must campaign perhaps we should be better occupied campaigning against the extermination of all life on earth, or for the alleviation of all suffering starting with human suffering. Should we not oppose worst things first?

From *The Eye of the Wind*

4

Badgers and the Animals of the Wiltshire Woods

I had an adventure the other night, and it has to do with the Wiltshire woods in general, and badgers in particular. I'm not an expert on badgers, or anything like that – in fact, up till last week my only experience of badgers was that I once caught one in an umbrella. Mind you, I don't suppose many people have caught a wild badger in an umbrella; I'd even hazard a guess that I was the only person who ever *had* caught a badger in an umbrella. But still, that doesn't make one a badger expert, and it was a long way from the Wiltshire woods.

It was up in the Bavarian Alps, where I used sometimes to go trout fishing in the summer – as a matter of fact, I think wild strawberries were as much of a lure to me as the trout. I was walking home across the valley one afternoon about tea-time, thinking of the lovely dish of strawberries I was just going to have, with a towel over one shoulder, because I'd been bathing, and my trout rod over the other, and hanging on my arm the umbrella which I usually took with me, because thunderstorms are a speciality of the Alps. Most hay fields in the Alps have a little wooden hut in them, where the hay is stored, and I came round the corner of one of these huts and saw a baby badger sitting right out in the open. Of course, as soon as he saw me he started to run back to his hole which was actually *under* the hut, and thinking it might be rather fun to have a closer look at him I jumped to head him off. We both had about ten yards to go, but as I went, I had a brainwave – I half-opened my umbrella and blocked the hole with it. The badger finished close second and he ran straight into the umbrella and got caught. Now, the chief thing about badgers is that they bite, and even this baby one looked pretty savage down in the bottom of my umbrella, so I wrapped my bathing towel round my hand and gave it to him to bite. He seized hold of it, and while he was busy biting it, I picked him up with the other hand and had a good look at him. Then I thought I'd like to show him to a friend of mine who'd come out to help me eat strawberries, and who was fishing farther down the river, so I carried him back to the inn. As we went he spent the time alternately biting my towel-padded hand, and rolling himself up into a sort of ball almost like a hedgehog.

After tea, we decided that he was much too old to be tamed, because although badgers make the most wonderful pets, you must start with them when they're very tiny. So we took him back to his home, and popped him down the hole under the hut, and then we listened. We heard the rumble as he ran down, and then we heard the most tremendous sound of grunts and squeaks. Evidently the poor little fellow was getting roundly scolded for being out so long.

So that was my only real experience of badgers up to last week, though I always wanted to know more about them, because there's no other English animal so comparatively common and yet so shrouded in mystery.

But I have a friend who knows quite a lot about them, and when she invited me to go and watch them with her in a wood in Wiltshire, of course I jumped at the opportunity. I spent two evenings there last week, and I want to tell you about them in detail, because to me they were both very exciting evenings.

On the first, we motored to the very top of the downs on the borders of Wiltshire and Berkshire. The best of the sunset was over, but there was still a marvellous golden glow in the windy sky, and on the downs there was quite a strong westerly wind blowing. From the crest of the down we walked over on to the north face where there was a little wood of hazels and ash trees, and in the middle of this we came to the badger's sett.

It was about nine o'clock when we got there, but it was already pretty dark inside the wood. On the steep slope we came to a little hollow and in it were half a dozen holes, and outside and around them were enormous heaps of white chalky soil which generations of badgers had excavated.

We lit a match to see which way the wind was blowing. Although it was westerly on top of the down, wind plays all kinds of funny tricks in woods, especially if they happen to be on a steep slope; the direction of the wind is very important when you're watching badgers, because their noses are much keener than their eyes.

We found out from our match exactly how the wind lay, and we worked round the down wind side of the sett to a pair of steps which my friend had specially put up sometime before so that the badgers could get used to it.

I climbed to the top of the steps and sat down on a rubber cushion I'd brought. I was about eight yards from the mouth of the main hole, and since badgers seldom look up, I was unlikely to be either seen or winded by them. My friend sat a little further away on the ground at the foot of a tree.

If you're going to watch for badgers, you have to know how to do one thing, and that is – sit still. I settled myself soon after nine o'clock, and I remained on top of that pair of steps until a quarter past eleven. I don't say that I never moved *at all* during those two hours, but I didn't move much: I just kept my eyes glued to the hole which was at the foot of a great ash tree.

Meanwhile I listened. Above the rustle of the leaves, I heard peewits calling from the downs, and once the stirring call of stone curlews. When it was still fairly light I watched a wood mouse foraging near the foot of my steps, but although I peered at the black mouth of the sett, the badgers never emerged, and at last, when it was really too dark to see anything at all in the depth of the wood, we decided to go home, and groped our way up the slope on to the open down, and back to the track which ran along its crest. There was a first quarter moon, and to the eastward, Mars shone a sort of rich coppery red. He was strangely ominous, I thought – Mars the planet, closer than he's been for nearly fifteen years. Mars the God of War. But, somehow, rabbits are reassuring things, and when we turned on the headlights of the car, a hundred of them scampered across the beam.

After that fruitless watch, we rather lost faith in those particular badgers, so during the next day we went to visit one or two other setts my friend knew of, and eventually in the afternoon we went to one where she'd watched them some years before but where they'd since been killed and their holes filled in. When we reached it though we found to our delight that there were two holes freshly opened, and bearing signs of evident occupation. The trunk of a tree near by showed recent scratches where the badgers had been sharpening their claws. Nearby, too, were lots of little scrapes where they had dug for the roots which form a staple part of their diet; and in the mouth of the hole were fresh fronds of bracken and leaves of dog mercury which badgers use for bedding. These leaves were scarcely withered, so they indicated quite certainly that the badgers had been there the night before. In great excitement, we tested the wind and chose our spot from which to watch and then since it was still much too early, we went back to the edge of the wood to await sunset.

As we walked along the ridges we found a lot of deer tracks, in the soft earth. From their size I think they must have been fallow deer. And then we came out on to the down, and the evening sun shone on to the great purple banks of rose bay. What herbaceous border can produce a more brilliant show than wild rose bay in July? The whole hillside was bathed in its brilliance. There was a little dew pond and we walked round it looking for tracks to see whether the badgers came to it to drink, but we didn't find any. The pond was pretty muddy, and every few seconds a newt would break surface and leave a bubble and then disappear again with a wiggle of his tail. There must have been hundreds of them – mostly common newts – though we did see one big crested newt.

At about half past eight, we collected our rubber cushions and field glasses and went back on tip-toe to our chosen watching place at the foot of a nut tree. On the way we lit a couple of matches to check the wind direction, and found that all was still well. Then we settled ourselves down to the really rather arduous task of sitting quite still.

We hadn't been there long before a wood-pigeon came to roost in a tree just above our heads, and we decided that he'd better be disturbed at once because he might have chosen to spot us and depart with a clatter just when the badgers were emerging. So we said, 'Off you go, pigeon,' and in a moment, he was gone.

About half an hour later, when it was still broad daylight, I happened to be staring full at the sett, when the first badger came out. I saw the top of his black and white stripey head appear, and my heart undoubtedly missed at least one beat, but of course, outwardly, I remained absolutely motionless. The badger was about twelve yards away and, after sniffing the air for a few seconds, he came right outside.

He seemed enormous, though he was probably a perfectly normal badger, but badgers sometimes weigh over thirty pounds, and this one looked about three and a half feet long. He stood just outside the sett mouth, and kept raising and dropping his head in a nervous way, and then sniffing on the ground, and then suddenly I saw another head emerging beside him. The second badger didn't stay up for long, and about a minute later there was a rumble and bump, and the two of them had disappeared again below ground. It seemed to us that they were very uneasy, and, indeed, it was still so light that I think it quite possible they may have seen us. Anyway it was another half hour before they came up again.

I'm afraid I can't begin to convey the terrific thrill of their sudden appearance. One moment, there was nothing, and the next – there was a badger, a dim dusky shape silhouetted against the pale green of the open, brackeny glade beyond the sett. This time the boar emerged first and sat down to have a really good scratch, which sort of echoed round the wood. Then the female came out too, and she had a good scratch. It was really an orgy of scratching. And then, a moment later, came a baby about half grown. The mother licked it and then both of them went down below again, leaving only the boar on top. He sniffed about but didn't venture more than a few feet from the sett. He seemed to me to be a bit suspicious still, and in a minute or two he disappeared again below ground.

By now it was pretty dark, and I kept the field glasses trained on the entrance of the sett, whilst we both very gingerly shifted on our rubber cushions, to uncramp ourselves, because we'd been sitting like statues for quite a while now.

We heard a woodcock doing his roding call in the distance, and a far-away peewit crying from the down, and then came the breathless moment when the badgers were there again.

Although it was so dark, I could see every detail of the boar's face as he re-appeared and started to sniff around again. He was twelve yards away and the glasses made him look an eighth of that distance – under five feet. Incidentally, my friend told me that she had once had a badger walk within three feet of her, so that afterwards she could reach out and touch the place where it had stood. But on this night we had no such luck, because their way seemed to be out into the brackeny glade beyond the sett. The two old badgers started off and then stopped and looked back, and then the baby came out and started after them. Immediately the old ones set off again, and the whole family trotted off up the wood for their night's foraging. We heard them moving at first, and then we gave them ten minutes to be out of earshot, before we got up and crept away.

The moon lit us down the ride, and we talked only in

whispers because the night was very still. When we came out on to the down, we found a glow worm shining away at the side of the path. We picked her up, and her tail shone on my hand, and then we left her in the grass. Rabbits rustled in the bracken, and the night was wonderfully still.

Well, that was my adventure, the sort of adventure that anyone might have if they took the trouble to sit still at a badger's sett. Of course, some people may not think it was very exciting, but, on the other hand, some may. I thought it was.

BBC National Programme broadcast, August 1939

5

The Introduction of Exotic Species

Successful species are constantly trying to extend their range, to colonise new areas. They are limited by the scope of their adaptability to a new environment, by their methods of locomotion, or by behavioural tradition. Thus seasonal food supply or winter temperature may limit the range of, say, the common chamaeleon in Europe; a narrow strait may prevent mice from colonising a new island; the family bond in wild geese may prevent them from resorting to pastures new because their parents did not lead them there. Some of these limitations have been overcome by the deliberate or accidental intervention of man. On the other hand, many species have extended their range by entirely natural accidents against which the odds must be very long indeed.

How many iguanas, lizards, geckos, tortoises, and snakes must have drifted down the Pacific rivers of South America on floating islands of weed before breeding stocks of each could establish themselves six hundred miles out from shore on the volcanic Galapagos Islands, newly risen from the sea? How many plant seeds had to drift out? How many land birds, like finches and mocking birds and hawks, had to be blown out to sea by offshore winds? These accidental natural introductions of animals to a group of originally lifeless volcanic islands helped Darwin to formulate his theory of evolution by natural selection, and led to an understanding of the principles of adaptive radiation.

The animals I have mentioned were introduced, so far as we know, without man's assistance. But the same islands are now also over-run with wild populations of donkeys, pigs, goats, dogs, cats, rats, and mice, all introduced on purpose or by accident by man himself. The list has been described as 'the wrecking crew', because of their effect on the natural ecosystem.

Introduced species have in many cases caused the extinction of indigenous species, because the native animals have not evolved to withstand the onslaught of new predators or to defeat new competitors for food, and cannot adapt themselves quickly enough to survive. In other cases, like the rabbits in Australia and the red deer in New Zealand, populations of introduced species have reached pest proportions with serious economic consequences.

These examples have led to a general conclusion among ecologists and conservationists that the deliberate or accidental introduction of any exotic species is highly dangerous, and basically to be condemned as ignorant, careless, and irresponsible. And yet every year thousands of animals are being translocated by a large number of people for the express purpose of releasing them into a new environment. Nowadays this is usually done in the name of sport – more precisely as part of the economics of sport-hunting and fishing. Less frequently it is done with the intention of providing biological control of some pest. In former times simpler motivations were at work – the desire of colonists to have familiar creatures, especially song birds, around them, or even just the desire to enrich a poor local fauna by making some colourful or attractive additions.

In the past, too, pigs, goats, and rabbits were deliberately released on uninhabited islands to provide a source of food for castaways. The pigs and goats on the Galapagos Islands are probably attributable to this source, and the feral pigs in particular are still a staple item of diet for the three thousand people living on the islands.

Accidental introductions have been no less disastrous than deliberate ones. Man cannot settle anywhere without bringing the rat, which has probably caused more extinctions of bird species than any other single factor attributable to him. When his two principal predator pets, dogs and cats, are added and become feral, as they almost always do in a habitat rich in native prey species, the position soon becomes critical for birds, small mammals, and small reptiles. The cats on Ascension Island, for example, are currently threatening the seabird colonies. Pigs have more often become feral accidentally than by deliberate human intention, and are particularly damaging to ground-nesting birds. They were certainly a major factor in the extinction of the dodo in Mauritius in about the year 1681. Pigs also eat the eggs of the various forms of giant tortoises in the Galapagos, and have been a major contributory cause of the extinction of at least two forms.

It is on the oceanic islands, which have so often evolved animal communities of particular variety and interest, that introduced animals have wrought so much havoc; and undoubtedly mammalian introductions have had the most serious consequences. In a very few cases bird introductions have been significantly inimical to man's interests; more often the effects of competition have been damaging to indigenous bird species, occasionally causing extinction. In the light of all these dire consequences, what should be the attitude of responsible people to the introduction by human agency of any exotic species of animals?

The conclusion must surely be that the unknown dangers of tampering with ecosystems in this way make *any* introduction hazardous and basically undesirable. This is particularly true of small discrete ecosystems, such as those of oceanic islands or isolated montane biotopes. Introduction of any mammals, and especially predatory species, is the most dangerous to the native fauna, insects probably the next most dangerous (with the possible exception of butterflies), then fish. Birds, which for sporting and aesthetic reasons are among the most popular animals for introduction, only rarely have a sweeping effect on the environment. It must be borne in mind that flight, combined with strong winds, has constantly led to the 'accidental' natural colonisation of new areas by birds. The recent spread of the cattle egret into the New World and the arrival of the collared dove in Britain are cases in point. A similar unsuccessful example was the presence in New Zealand of a small population of the Australian white-eyed pochard for a limited period in the early part of this century.

The successful introduction into Britain of such birds as the little owl, the Canada goose, the mandarin duck, the pheasant, the red-legged partridge, and most recently the ruddy duck do not so far seem to have adversely affected the environment or the indigenous avifauna. In other areas introduced species have been less welcome – for example, the starling in North America and elsewhere, the Indian mynah in Hawaii and other Pacific islands, the house sparrow, etc.

In New Zealand the desire of colonists for familiar creatures from the homeland led to the establishment of a network of 'acclimatisation societies', vieing with each other to introduce exotic species. This has had a very significant effect on the endemic avifauna. Deer of a number of species have greatly altered the nature of the forests; Australian possums have enormously reduced the native birds. The wetland ecology has been changed by the addition of Australian black swans and Canada geese, both now present in tens of thousands. The northern hemisphere mallard has crossed extensively with the New Zealand grey duck, producing a rather ugly hybrid which is now widespread.

Although these results are to be deplored, it is still open to argument that justification may exist on aesthetic grounds for the introduction of colourful or melodious birds into a dull native avifauna. Such argument might be extended to include butterflies and perhaps some small reptiles and amphibia. A second justification might be established for the introduction of biological controls for pests as an alternative to chemical pesticides, though it should not be forgotten that the introduction of the mongoose to control rats has been unsuccessful and singularly disastrous in many places, notably in Hawaii.

Introduction for biological control might consist of predatory or parasitic invertebrates, though even then the results may well turn out to be unexpected and embarrassing, and clearly the technique should only be used sparingly, and then only when it offers a promising alternative to significant human hardship or suffering. A third possible justification might be found for translocating a species seriously and irrevocably threatened in its own range to a new area outside its own range. This was the intention (at least in part) for the introduction fifty-eight years ago of the greater bird of paradise into Little Tobago.

Another aspect of the subject is the re-introduction of animals into areas which they formerly inhabited, but from which they have been exterminated. A recent proposal suggests that the wolf and the bear might be reintroduced to Britain to control over-population of deer in some areas. This is only practical if the original causes of extermination have been removed. In this case, conflict with man over depredations of domestic livestock was probably the principal original cause, and it seems likely that the same conflict would bring the reintroduction experiment to nought. It is questionable whether a state of opinion has yet been reached in which man will be ready to accept material loss by depredation for the advantages of ecological balance and an enriched fauna for his aesthetic enjoyment, although in the long run such a state of affairs might well be attained in developed countries. The idea that some small financial sacrifice may be necessary and desirable in the name of aesthetic enlightenment and civilisation is gaining ground in some countries quite rapidly.

When a form has become totally extinct it could be worth considering the introduction of a closely related form which might be adaptable enough to become established. For example, another race of the North American prairie chicken might be introduced into the former range of the now extinct heath hen, provided that a viable amount of suitable habitat could be set aside for the purpose.

The tendency of man to wish to interfere with the natural distribution of animals is not always associated with practical or material motivation. There seems to be an element of curiosity to see whether an artificial introduction 'will work', and a sense of challenge to make the attempt a success.

We may regret man's tendency to 'play God' in this way, and we may feel that the purist approach which deplores any introduction whatever is the more responsible view. We should not forget that whichever attitude finally prevails, man has already greatly modified the distribution of animals across the earth, and will certainly continue to do so. We should also remember that man and his attitudes are more than ever an essential ingredient of natural evolution, because man is still an integral part of nature.

Address to IUCN Technical Meeting, Lucerne, 1966

6

Food Production and Conservation

In a world where two thirds of the human population are still hungry and vast numbers still die of starvation or diseases of malnutrition, a visitor from outer space would have difficulty in understanding why a large proportion of the remaining third suffers in various ways from over-eating. The same visitor would have difficulty in understanding how a species that is fully conversant with the principles of population dynamics is unable to apply that knowledge to its own survival and prosperity.

The curve of human population rises steeply; so too does the curve of food supply, but not steeply enough. Far from the gap between them closing, it continues to widen. Demand exceeds supply by more each year, not less. In increasing food production we are not even keeping up with the increasing population. The long-term answer must lie in the power of mankind to control his destiny by controlling his population – a forlorn hope, many people think, but as a practising optimist I disagree. As an animal, man is more destructive than any other, but he is also more imaginative and creative and ingenious than any other – and he has evolved a conscience. Throughout his history he has been surprisingly good at removing obstacles, at eliminating things that were clearly disadvantageous. As soon as over-population becomes clearly and *universally* disadvantageous, the problem that now looks so insoluble will be solved.

Meanwhile, back on the ranch ... food production must somehow be stepped up to tide over the period before man discovers that quality in life is a better aim than quantity, that fewer people with higher standards of life are better than more with lower standards. If the lesson is not learned, human beings will soon be living in much the same conditions as factory-farmed animals.

But though I am certain the lesson will eventually be learned, we have still got to step up food production, and this is where Elspeth Huxley's book *Brave New Victuals* comes in. It looks with acute perception into the future and it deals in remarkable detail with the present trends that have caused so much concern to many, myself among them. It deals also with the reasons for that concern. Time snaps at our heels. There is no time to lose, we realise; no time for long years of research which may yield results that may be unexpected and even contradict some of our accepted beliefs. As the human flood spreads over the countryside, robbing the farmer and stockman of land needed for his crops and livestock to cover it with habitations, roads, schools, factories, recreation grounds, and power stations, so we must step up output from the land which remains. So we turn to those intensive forms of food production many of us instinctively dislike – battery hens, barley beef, indoor veal, sweat-box pigs, and so on. We cannot afford to sit back and let the pests, weeds, and diseases, formerly kept in check by traditional methods of husbandry, take a mounting toll as the opportunities offered them by modern concentrated methods of production expand. So we rush into the indiscriminate use of pesticides which kill not only the pests, but all the other insects and the creatures which feed on them, and the other creatures which in turn feed on *them*, until the food chain brings the poisons to us – not, as we are constantly assured, in harmful quantities, but in measurable amounts. To steepen that food production curve these risks may have to be taken, even though research on side effects is incomplete. If the effects of pesticides on predatory birds are a true indication, then we may reach a point where man (the most predatory of all predatory animals) begins to suffer from a similar sterility. From an over-population point of view it could be argued that such an accidental side effect would be quite beneficial. But if man's fecundity is to be limited, surely it should not be left to a blind miscalculation which at the same time kills or sterilises so many other creatures? Surely man will demand more control of his destiny than this?

The technology of our world accelerates as 'progress' rushes onwards. Only the mind of man and the speed at which it can be changed or modified or moulded to new ideas remain sadly the same, or almost the same, as ever. The adaptability of the human animal progresses inexorably at the speed of evolution by natural selection, which now includes the special social selection operated by man himself. It may take one tenth of the time that it took ten years ago to develop an electronic invention, but it still takes just as long as ever to adjust the human mind to a wholly new idea. So technology outstrips our capacity to control it and use it for the enrichment of human life. Instead, we are in danger of being used by it, and driven over the precipice like Gadarene swine in the grip of a spirit we do not understand.

Moses took the children of Israel into the wilderness for forty years. Evolution might have brought that figure down to thirty-nine years, some four thousand years later. But now we can neither afford to think in terms of thousands of years, nor even of forty. The dangers are immediate, as the gap between the under- and the over-fed widens, and through the spread of education the underfed and overcrowded come to realise that their condition is not due to the will of God or the decrees of fate, but rather to the inability of man to control his own destiny and organise his own resources. Then they will demand changes, not in forty years, but here and now.

This is the background to Mrs Huxley's inquiry. Hers is a very important book. It is about the survival of the human race, and it deals realistically not only with facts and figures, but also with human emotions – a part of the equation which planners often forget. We all like to think we are logical and rational, and motivated by what is scientifically sensible – but we all in varying degree have emotions, and bees in our bonnets too, stemming from our personal dispositions and experiences; without them we should be incomplete human beings and our decisions would be out of balance.

We are perpetually being faced with decisions – the 'don't knows' are usually the smallest category in any popular poll – even though there may seem to be little enough chance of influencing the course of events by the stand we make. At least we should be informed. That is the service offered to us by this book. Elspeth Huxley has given us the facts, as she set out to do. The judgements are up to us to make.

Foreword to *Brave New Victuals*

7

Search for Salvadori's Duck

Wednesday 12 December 1956. Off to New Guinea at last. My wife Philippa, BBC cameraman Charles Lagus, and I embarked in a DC4 with hordes of schoolchildren returning home for Christmas. We left Sydney at 4.15 a.m., had breakfast at Brisbane, and flew north along the Queensland coast. We had our first view of the coral atolls of the Barrier Reef, green in the shallows, so green that after a while the little cumulus clouds looked pink. And then in the afternoon, with a deal of turbulence at low altitude, we landed at Port Moresby. This is in arid savannah country with sparse eucalypt bush. We were met by the representative of the Hallstrom Trust – Barry Osborne – who was helpful and pleasant and saw us into the DC3 which was to take us on to Lae. It was a freighter and we sat on benches along each side.

We flew through thin and wispy cumulus, over steep wooded ridges and swamps. It was green and lush and jungly. And at 5.00 in the evening we circled round Lae, seeing how the town has surrounded the airstrip, and how the waters from the Markham River laden with mud sweep round the north shore of the Huon Gulf (in spite of the clean water of the little river at Lae itself) so that a Japanese ship beached in the war is now so silted up that it has become entirely surrounded by land. Only from the air can the new bulge in the coastline be detected.

At Lae it was hot, unbearably hot, being evening. We were met by John Cunliffe – among other things Veterinary Officer – and by Mrs Taylor of the Department of Agriculture. It seems that Sir Edward Hallstrom had given no prior notice of our arrival to anyone and his telephone message by radio had become considerably garbled. It appears that they were expecting two of us – me and Director Severn [a reference to The Severn Wildfowl Trust, now the Wildfowl Trust]. Where, I was asked, was Director Severn. 'That's me!' 'Oh, then where is Mr Scott?'

As we drove to our hotel I was suddenly excited by the whole prospect. This was New Guinea. It looked strange and exotic. It was superbly tropical. It was the naturalists' paradise, as I had always read, and I suddenly felt sure it was going to live up to its reputation. Besides, it was the home of the mysterious Salvadori's duck, which I should soon be seeing for the very first time. It was dark by the time we were established in our rooms, but I could see at once that the population of geckos was higher than anywhere else I had ever seen. Newly hatched young and adults were all over the ceilings and around the lamps. The passages were full of them. Every few minutes we heard the delightful little clicking call, *Tche-tche-tche-tche-tche*, not unlike the noise one makes to persuade a horse to go faster. The call gets softer and the repetitions quicker at the end.

Friday 14 December. We left early for the flight to Nondugl. At the airport Charles had Customs difficulties. It appears that we had not been properly cleared on Wednesday evening, and the Customs officer now wanted a deposit of £250 on the value of Charles's cameras and film. Honour was finally satisfied with the gloriously illogical arrangement that Charles was to leave one half of his letter of credit as guarantee of the deposit! Then we boarded our DC3 freighter with a great pile of furniture and other freight in the centre and a bench of seats along one side only. There were six passengers including us.

At one of our stops, Hagen (pronounced almost to rhyme with bargain), the District Commissioner named Skinner came up and introduced himself. He told us his son had shot two Salvadori's ducks with one shot – at eight years old. Skinner called over and introduced Ned Blood as the most knowedgeable person on ducks. He had been at Nondugl for some years. He said the only other ducks in the Wahgi Valley were black ducks, but that over on the Sepik there were 'lots of ducks including the harlequin duck'. Just then the plane was about to take off, so I never had a chance to find out what the New Guinea harlequin duck might be.

And so we flew to Nondugl. I liked Nondugl the moment we were down. It was near enough to the northern foothills to have dominating hill features as a backcloth and somehow the broken country and the trees made the whole landscape just right. Nondugl was at once apparent as a lovely place. Frank Pemble Smith arrived on a motor bike, and soon after a tractor and trailer arrived with Fred Shaw-Mayer to greet us. These were two very different men. 'Pem', thick-set, self-assured, a pioneer, an Empire-builder. Fred, tall, quiet, a scientific collector and ornithologist, now in his late sixties, burned dark by the sun, slow of speech, courteous of manner.

Fred rode up with us in the trailer and we came to the homestead and were greeted by Sue Pemble Smith – attractive and friendly, with two small boys, Peter and Timmy. Once installed, with a sandwich lunch inside us (and pineapple grown locally) we went down to 'the Fauna Section' – which is Fred's zoo. He showed us round and it was an utterly delightful afternoon. The zoo (or should it better be called the sanctuary?) covers about 10 acres. It consists of three ponds formed by dams across a shallow cleft. Around these are lawns and flower beds and flowering shrubs and bamboos, and dotted about are eight or nine aviaries, beautifully planted up and somehow fitting most appropriately into the whole landscape. The collection consists of about two hundred birds of paradise, some bower birds, parrots, three cassowaries, and twenty Salvadori's ducks and a dozen or more tree kangaroos of three different kinds.

It was not unnatural, perhaps, that we could hardly wait to see the Salvadori's ducks. Fred took us into the enclosure round the top pond and there at once was a pair of them about 15 yards away at the edge of some sedges. The bird was more striking than I had expected. The yellowish, flesh-coloured bill with no black tip was prominent and there was quite a sharp dividing line between the dark brown head and the pale, creamy-yellow ground colour of the spotted breast. The barring of back and flanks, however, was less precise and neat than I had painted in my illustration for *The Waterfowl of the World*. Nevertheless it was quite a handsome bird. More surprising were its movements. It swam with a rapid jerky backwards-and-forwards movement of the head, caused apparently by nervousness, or watchfulness, and the tail, which is very long, was carried at times almost vertically.

After a while we saw the display. It was the only true courtship display behaviour we saw, though later we saw 'dashing and diving'. The male's head is stretched up and moved only slightly in unison with a whistle. The whole of this is reminiscent of the shelducks. The female meanwhile pumps her head up and down violently, emitting a series of croaking barks which are clearly analogous to the mallard's quack.

In the next few days we had ample opportunity to become familiar with the bird. In feeding it frequently 'up-ended' like any other dabbling duck, but it dived much more readily and went under from a 'low-in-the-water' posture without making a ripple. It is obviously a very accomplished diver. It is also of skulking habit and largely nocturnal, as Fred told me of various occasions when he had seen them about the pens at night. Later I had a chance to handle three live birds – two males and one female. The female's head was much paler on the crown and forehead, and more streaked. The wings are quite short, but the speculum is very well marked, though the green is only on the inner half. The line of the white tips of the secondaries is extended to the tips of the inner primaries.

From all that we saw of the bird I have no doubt that it is a dabbling duck and should correctly be in the tribe Anatini, though whether it should be in the Genus *Anas* is more doubtful. It is certainly not just another teal like the Cape teal or Chile teal. It is quite highly specialised for living in mountain streams; the long tail and the jerky head movements are reminiscent of the torrent duck. The eye is rather farther forward than in most dabbling ducks, suggesting the requirement for increased binocular vision in order to catch living prey. Fred thinks that tadpoles form the basis of their diet in the wild state. He feeds his ducks with tadpoles during most of the year, but in the wet season, when they are harder to get, he augments this with meat – sometimes tripe.

Saturday 15 December. Pem took us for a drive in the Land-Rover round the estate. We went first up the slope to the site of a new training college which is to be built. From here we looked down the easy sloping side of the valley across pastures dotted with casuarina trees. The Hallstrom Trust ground is a narrow strip half a mile wide and two miles long. At right angles to it farther up the Wahgi Valley, and detached by a couple of miles, is another narrow strip of about 800 acres. These lands are bounded by native holdings. The population is very large and yet one is hardly aware of it. Nowhere can a dense human population blend more perfectly with its environment. The little low grass houses are hidden away in the shade, and only occasional glimpses of banana trees, or sugar cane, or the squared beds of sweet potato on the hillside show that the landscape is subject to man's influence.

After lunch we returned to the pond in the zoo. It was poor light with intermittent rain, so we occupied the time building a hide beside the water close to where the ducks seemed to be principally loafing. Fred has now bred them twice, rearing three and two from two clutches of three eggs each. He has twenty altogether, of which fourteen are on the main pond, four more recently captured in a very small pen with a little pond, and a pair on a lower pond, which were in full moult with no flight feathers. One pair bred in January and another in July; it is possible that it was the same pair, though it seems to me unlikely. Only four of the fourteen on the big pond were in evidence. These four seemed to get on quite well together and occasionally stimulated one pair to display.

Sunday 16 December. Charles got up early and recorded the howling of the wild dogs. The pair of New Guinea wild dogs kept in the zoo are the most delightful animals; they look like a cross between a fox and a dingo, and I think they might well become a highly successful new breed, if properly introduced. Phil and I were also up early, and filming Salvadori's ducks on the lower pond before breakfast.

They behaved fairly well, although they usually retired into the shade before displaying. Later, in the afternoon, Phil took some stills from the same hide, while Charles filmed Gouwra crowned pigeons. Most afternoons it had begun to rain by about three, though some days it had held of even until five.

Monday 17 December. The day's objective was to see Salvadori's ducks in the wild. For this we were to go up the Gonigel, a river flowing past Nondugl from the north.

We set off in the Land-Rover and drove as far as the road went – a timber road cut for getting out logs. Our guides were Aroba, or Roba – the Pemble Smiths' house boy – and two young lads who worked for Fred Shaw-Mayer, called Aggis and Kahboon, who were supposed to have located some ducks the day before but hadn't found any. It was still grey and cloudy when we got to the end of the road, and indeed it remained so all day.

The road ended at the river and we were supposed to ford at

this point, but the first of our band of a dozen or more natives who had assembled found the crossing too deep. So, having removed shoes and socks, it was decided we should try a few yards upstream. This few yards developed into quite a climb along a cliff face, over a ridge, through a pandanus swamp and eventually to the river about a quarter of a mile farther up. We squelched in our bare feet through the mud pools along a scarcely detectable path. At last a crossing place was detected and Charles filmed us all wading thigh-deep in the white water, making a chain across the river. Once across I filmed Charles following.

On the far side we put our shoes and socks on again and started up the side of the river again. Soon, however, we joined a more major path, leaving Roba to follow the river and report if any ducks were seen. We met natives along this path, which wound through high timber, past native grass huts and gardens, up into the moss forest. Here mosses and lichens grew on many of the trees. There was surprisingly little in the way of animal life. The only bird I saw to identify was a yellow-faced honey eater.

It was pleasant climbing and the boys were carrying all our camera equipment, coats, and so on. By now we were quite hot, although the valley was still overcast. But the cloud had been receding uphill ahead of us, so that we were never walking in mist. Eventually we reached a primitive foot-bridge across the river and not long afterwards the stream divided. At this point Pem suggested that we stop for lunch, sending the small boys on ahead to find ducks.

It was an excellent lunch of grilled chops, tomato sauce, excellent homemade biscuits, banana, and pineapple. During, before, and after lunch various small animals were seen and I filmed two colour phases of a fine balsam – *Impatiens sp.* – one of which was purplish pink and the other orange scarlet. A rather nice beetle with a yellow tail was brought in by the boys and I watched a very beautiful red, yellow, and blue tiger moth trying to find somewhere to lay an egg. She landed for a moment on my shorts. After laying an egg under a very large leaf, the culmination of a couple of minutes of hovering, the insect swept up and away with unexpected speed.

The boys returned after lunch, standing round the fire to keep warm. They had seen no ducks. They now said that it was no good up here in the rainy season, the river was too swift, the ducks had all gone down to the swamps along the Wahgi. Yes, Roba said, he knew the duck well. It was called 'Koorang' or 'Koorund'. We got them all to say it, and finally it was 'Toolund'. I did a drawing for them and they all agreed. There were two kinds down on the Wahgi – the big ones which were dark and the little ones which were paler. If we were to go down to the Wahgi he would show us that he was not telling lies. All this was communicated in New Guinea Pidgin, which is just understandable when Pem talks it, but quite incomprehensible when spoken by the natives.

After lunch it seemed that nothing remained but to turn back, especially as the rain might soon be expected to start. Actually, it held off as we made our way back down the path, filming as we went. We stayed on the main paths, which finally led down to the ford opposite the Land-Rover. When we got there Pem and Charles were already across. It appeared that Pem had been swept off a rock and had fallen in completely, and that Charles had filmed the whole thing. They suggested that we should go a few hundreds yards farther down stream where there was a bridge. This we did and were met there later by the Land-Rover.

Charles was wading about in the stream below to film the crossing of the bridge, which consisted of three pandanus trunks lashed together with vines, with a single branch like the prong of a fork as a hand rest for balancing. Once safely across the bridge we were soon embarked in the Land-Rover with a mass of natives piled in the back with Charles and the cameras, and on our way down the valley to Nondugl.

The rain still held off, so we went down to the zoo again and filmed Fred and Phil and I going to look at Salvadori's ducks. We had no sooner finished than it came on to rain hard, and we had to run for it. And during the night it rained and it rained, so that the next morning . . .

Tuesday 18 December. The cutting on the road was blocked by a landslide and our trip to the Wahgi had to be postponed. Instead we filmed and recorded the yodelling call by which news is passed round the valley.

Late in the afternoon a most beautiful rain storm with back lighting appeared over at Minj across the valley. That evening Fred Shaw-Mayer came in for dinner. He told wonderful stories of his early travels and collecting trips. After dinner I got him to give me the names of some of the birds we had seen and I went on picking his brains until far too late, for which I was in trouble from Phil at bed-time.

Wednesday 19 December. Pem and Phil and I made the trip to the Wahgi in the Land-Rover, while Charles stayed behind to film birds of paradise. It was a very hot sunny morning. We went down the main road for a few miles and struck along a ridge to the edge of the last ledge above the river. Here we looked across the bends of the meandering river, *café-au-lait*-coloured after the rains. The river flowed fast, but there were some swamps which were the remains of ox-bows, old river channels. One of these, which was almost overgrown, was immediately below us and Roba suggested we should go down past a little homestead and its pigs and see if there were any ducks. Another ox-bow on the far side of the river had an egret beside it – *Egretta garzetta nigripes* – and a couple of black dots which were almost certainly black ducks. What would there be in the much thicker swamp below us?

Accompanied by two little girls we got down to the edge of the swamp, but could find practically no vantage point from which we could see what open water there was. It became clear at a fairly early stage that there was no great future in this particular swamp. A large bird flying over the river in the distance was pronounced to be a duck, but was in fact a cormorant. The only other birds we saw were a pair of rather pretty little shrike with a red back. This was *Larinus schack stresemanni*. There was also a pale, harrier-like hawk with black shoulders of which we saw two or three during the day – it is *Elanus coeruleus wahgiensis*. The 15-foot-high reeds were full of little brown finches, the males with smart black heads – the Wahgi munia, *Lonchura spectabilis wahgiensis*. Roba and a local who had joined us were sent to look for ducks on the next bit of swamp. It was half an hour later when our spirits were a bit low that conversational quacking began in the reeds quite close by. This was evidently a black duck, but it showed that there were ducks about.

We were now about 1,500 feet below Nondugl (5,500 feet) and it was becoming very hot indeed. The prospect of a 500-foot steep climb back to the Land-Rover was not inviting, but we decided to make a start, having persuaded the little girls to call to Roba. Halfway up the hill we stopped to rest and were joined first by a very attractive fifteen-year-old girl and then by a charming old man – father of all the children – who came up making a most peculiar noise like *Eeee-eeee-eeee* which was evidently an expression of extreme pleasure at meeting us. He was quite delightful. Roba explained his joy with the phrase: 'Big fella master, big fella Mrs'. It is a commentary on the size of these natives that Phil should be regarded by them as 'Big fella Mrs'.

We decided to have one more try at the Wahgi swamps by going round and over the bridge and then coming back on the other side so as to overlook the pool on which we had seen the pair of black ducks. So, extremely hot and deliquescent, we got back to the vehicle and set off upstream towards the bridge. One of the nicest features of the roads is that they are planted with flowering shrubs, especially plants with red leaves. Among them is a pink flowered balsam with red leaves. The Wahgi bridge is a suspension bridge on wire cable which produces a wonderful wave action as the vehicle advances across it. We called on a planter who lives near the bridge but he was away at Minj. However, Pem raided his larder and we ate pineapples after we had examined a pond just below the house, on which there were no ducks, but which looked as though there might have been. It was in this garden that we saw a tiny dark quail, *Turnix maculosa giluwensis*. Then on to Minj where we turned back down a track leading to the river again.

But the trouble with the track was pig fences. The first one we came to had two natives beside it, and Pem decided we should dismantle it; the natives were a bit sullen about it but eventually they helped Roba with the job. Two more fences were dismantled and then we came to a stream with a pandanus trunk bridge and the car could go no farther.

Then began an endless walk along a black, single-track foot-path between hedges of *Crotaleria anagyroides*, a shrubby yellow leguminous plant with leaves rather like laburnum. The flowers are about the same size too, but on a spike instead of hanging. Frequently we thought we were getting near the river only to find that another long stretch of level path opened out in front of us. At one place where there was a dip in the path we came upon a most colourful muster of a dozen or more natives with fine head plumes and looking magnificent against the mountains at the crest of the rise. We wondered if this was their everyday wear, or had the bush telegraph given warning of our coming? Certainly we had heard no yodelling shouts. In the bushes beside the path we came upon some very beautiful little birds with sickle-shaped bills. They were very small – about the size of a bluetit and mostly blackish, except for the head and throat which were brilliant scarlet. These were *Myzomela adolphinae*.

At last we came within sight of the three trees which we had marked from across the valley, and a few minutes later we were climbing the last ridge to look over at the swamp below.

At first we could see no birds on the water at all. The pool was two or three hundred feet below us and as many yards away from us. But we sat down with the glasses and quite soon I saw a black duck, then another and later a third. Over on a more overgrown part of the marsh were two swamp hens, *Porphyrio sp.*, and beyond on a small stream two more black ducks swam into the open. A little later the egret which had been there in the morning reappeared. Flitting over the water were several Willy wagtails and screaming down among the trees on the slope were some little parrots – probably *Neopsittacus pullicauda*.

As we sat there an extremely exciting moment arrived. From the rushes on the right of the pool swam out two small ducks. Any small ducks there should have been Salvadori's but these quite certainly were not. These were grey teal, *Anas gibberifrons*. As they swam out across the water they passed behind a tree in the topmost branches of which sat a bird looking large and prominent and reddish. This was a male red bird of paradise (*Paradisaea apoda salvadarii*) – though not in full colour. As we identified him the grey teal swam immediately behind him.

It was one of those great moments in ornithology when two exciting things happen at once and there is not enough time to think of both. After a while the red bird flew away and the grey teal went ashore on the left-hand bank of the pool. There they preened and finally went to sleep. They seemed to be stained reddish on the breast, which was very pale. It was disappointing that they were not Salvadori's ducks, but the grey teal has only been recorded once from the valley. No native had ever reported *three* kinds of ducks in the Wahgi Valley. Evidently they did not differentiate between *waigeuensis* [Salvadori's duck] and *gibberifrons*. This immediately indicated that the swamp habitat for Salvadori's duck was an error and at once the whole story began to make more sense. Salvadori's duck is a bird of highland streams and lakes. It is analogous in habit with the South American torrent ducks, and the long tail is an adaptation for the swift mountain streams. It has never lived in the Wahgi swamps nor is it likely to occur below 5,000 feet.

And so we decided to set off on the homeward walk, especially as it was already about 3.00 p.m. and the rain clouds were gathering, with thunder to the north. The walk back lasted forty minutes. In due course we came back to Minj, and later, through rain to Nondugl.

Thursday 20 December. This was the day of the 'sing-sing'. Soon the natives were assembling in full regalia for the dancing which was to be laid on especially for us. There is no doubt that the head-dresses made largely from the plumes of birds of paradise are superbly colourful, and there is a striking variety in the form and structure of each head-dress. At least five species of paradise birds were represented. But the actual performance of the dances was a little disappointing. It was under the direction of the present chief who was dressed in a dirty shirt and shorts with an American peaked sports cap with a transparent green eyeshade under which, across his forehead, he wore the medal of chieftainship issued by the Administration. He flitted about, acting each part, like a producer at a dress rehearsal. Most of the dances were mock charges, but it seems that they never really got worked up and it was all a bit hot and a bit boring for them. However, in spite of that, we got what should be some very fine pictures. We filmed from various angles, I at one time from a hollow dug for a banana tree, Charles from a huge timber platform which was brought up on a tractor about halfway through the proceedings. At times the boys jiggled up and down in time with a drum carried by one of them which was shaped like an elongated egg-timer and covered with the skin of a monitor.

By lunch it seemed that we had had all we were going to get. Pem had promised them some money and when we got back to his office he gave me a bag of silver and through the window, as at a school prize-giving, I presented each one with five shillings. Only one of them, a youngish, rather good-looking one who had been rather sullen about the whole affair, appeared to check how much was being given and then to refuse to come up for his – on the ground, one supposed, that it was not enough. But when Pem asked him if he had had his yet, he immediately came up to get the money.

And so we assisted, if only in a small way, in the degradation towards the white man's level of the Awahga natives. There is nothing that a civilisation can do to make the New Guinea native remotely more happy than he was before his existence was discovered by white explorers. In the Wahgi Valley one hears of all kinds of praiseworthy ideals which attempt to convey that the enlightened colonisers are bringing untold benefits to the poor ignorant savages. But there is only one way in which we could play absolutely fair with the New Guinea natives and that would be by declaring huge areas completely out of bounds to any white man at all, and leaving them entirely alone. I have no doubt whatever that the Awahgas are the better without us.

But as this would be putting the clock back, and as the grasping cupidity of man is such that for all his loudly proclaimed ideals he cannot resist exploiting new lands without adequate thought for their original inhabitants, the natives of these lovely highlands will be spoiled, debauched, and degraded by the process of what we call 'civilisation'.

From *Animals* magazine

8

Skin-Diving: a New World

Sunday 23 December to Tuesday 25 December 1956. For a part of these three days at Cairns, on Australia's Queensland coast, I have been in a new world. Nothing I have done in natural history in all my life has stirred me quite so sharply as my first experience of skin-diving on a coral reef. Konrad Lorenz said, when I saw him in Bavaria in September, that this was one thing I must do before I die – and now I have done it, or rather started to do it. The dramatic threshold which is crossed as soon as one first puts one's face mask below the surface is, to a naturalist, nothing less than staggering in its impact. Much has been written about the scarcely explored new continent of the ocean; I have read these descriptions in the works of Cousteau and Diolé, yet I was unprepared for the visionary revelation when I first saw the real thing. I must try to describe it chronologically and in some detail, but the effect on my mind is still rather kaleidoscopic and bewildering.

First it should be explained that the adventure falls into four chapters and an appendix – by the way, I have no hesitation in using the word 'adventure', for this type of swimming cannot fail to be high adventure, and nothing less, for any naturalist, indeed, for any imaginative person who has never done it before. The four chapters were four separate dives – two on the first day and one each on the next two. The appendix is two visits to the Underwater Observatory on Green Island. As befits such a sequence, each chapter was a little bit more exciting and more moving than the one before. The final effect was overwhelming, so that in spite of trials and tribulations with ill-designed equipment, and intense discomforts arising therefrom, I cannot see how I can escape from its lure. I am already an addict and I have not yet used an aqualung.

Chapter One began when Hughie Reid came to fetch us in a beat-up Vanguard estate car and drove us north to a beach to pick up a boat which we put on top of the car; then we drove yet further north to Trinity beach, off which lay Double Island.

We launched the boat, which was called *Radish II*, and Hughie rowed us across the edge of the reef. The water was unusually muddy because of the recent cyclone which had struck further down the coast but whose repercussions had set up strong winds in this area. Charles, who has done quite a lot of skin-diving, knew that it should be better than this, but when I slipped over the side of the boat and adjusted my mask and snorkel I looked vertically down at just visible coral about 5 or 6 feet below me, and saw fish going comfortably about their business among it. In those first few minutes I saw a dozen kinds of fish, I was shown the blue and purple edges of the mantle of the clam – a curly line of colour which narrowed in sudden jerks at the approach of my shadow, and I had my first chance to differentiate between hard coral and soft coral. There were too many different kinds of coral for them to be catalogued in the mind: each and every one seemed different from its neighbour. I borrowed Charles's New Guinea arrow, which he was using as a probe, and touched some of the soft coral. In one large 'bush' of flesh-coloured soft coral, when I touched a twig the whole branch drew in its tentacles and the fingers looked darker and thinner. There were about four branches in the bush, which I supposed to be four corporate coral 'animals'.

Already I was having some difficulty with my mask and mouthpiece of my snorkel. Charles and Hughie had brought up mushroom coral (*Fungia*) and other kinds, one piece with a brittle star on it. They had also brought little orange crabs with big claws, a sea-cucumber, and a little flat shrimp that had emerged from another coral. We went ashore on Double Island. It was blazing hot and one could hardly walk on the beach. There were little mouse holes on the beach, and these were apparently the homes of sand crabs. I dug one out. He was surprisingly large and handsome with eyes on long pointed stalks which could be carried horizontally or raised comically one at a time.

After lunch we decided to dive on the other side of the point. This was Chapter Two of my new adventure. There was not much coral here but there were rocks along the steep shore. The water was much clearer – so much so that I could see the fish quite a long way ahead, say 20 feet or more. This gave much more of a side view and less of a top view of the fish than I had had before lunch. But the absence of coral was more than compensated for by the fish which were more in numbers and in kinds than earlier.

It was here that I saw my first chaetodont, a square fish with a long snout and a beautiful zig-zag pattern – more often called butterfly fish. There seem to be a lot of different kinds – slightly different. To what extent this indicates different species I have still to discover. Maybe there is a sex difference or polymorphism or just plain individual variation. Growing on the rocks here was a curious shell with corrugated edges and clustering so as almost to cover some of the large stones. There must have been at least twenty different kinds of fish round these rocks. The largest and most exciting I saw was a parrot fish – related, I think, to wrasse. I saw two together, one larger and more brightly coloured than the other. The mouth is parrot-like, the colours are green, blue, and red and the bright blue pectoral fins flap up and down just like a parrot in flight. In fact, it is a very good vernacular name.

I was having quite a lot of trouble with the rubber mouthpiece of my snorkel, but so intense was the enjoyment of what

I was watching that I hardly noticed that my lips were being badly damaged by it. I swam between stratified, near-vertical sheets of rock – I was in the nave of a leaning church. Although there was a slight swell I did not feel in serious danger of being swept onto them. The only times that I scratched my fingers was when I tried to hold on to a rock in order to adjust my mask and snort. I chased a small fish up a rock passage and finally he went to ground under a stone. Later I found a shoal of tiny sardine-type fish, and showed them to Charles. We cornered them against the shore and they dashed past us. In the shallows under rocks were quite large dark blennies. The last excitement was a shoal of about twenty round fish like vertical dinner plates, about 8 or 10 inches across, which stayed in a decorative and sociable cluster about a foot above the rocks. They were brown and marbled and their sociability was vastly impressive.

Soon after this I decided to go ashore. Only the intense pain in my lips made it possible for me to stop cruising among the fish. The greater clarity of the water on this side of the island had opened up new vistas. A very curious feature of swimming with a snort is that although one's eyes are only an inch or two below the surface, there is no feeling of being on the edge of the world one sees. On the contrary, one feels very much a part of the scene, and of course one *is* in among the fish, for one's hand reaches out towards them, and some are hardly scared by it and move quietly only a foot or so in front; one's feet sink down to the rock, or the coral, below, and the nearest fishes dive for cover into the crannies. It is as new a sensation as gliding or skating, and with it all is the almost agonising thought to a naturalist – of how much there is to learn before one can begin to know or use all that one sees. And above all there is the incredible beauty of it.

Monday 24 December. Christmas Eve, and an early start. We were collected by Mr Wire, Secretary of the Cairns Harbour Board, and conveyed to the harbour where a fine launch was waiting and soon we were off to the reef.

It took about three and a half hours to get to Michaelmas Cay, past Upola Cay. The island is very small, two acres of rough grass and marram above a dazzling white beach. After filming the huge tern colony the decks were cleared for Chapter Three of my underwater adventure.

Putting on the mask and snorkel was agonisingly painful. My mouth was now swollen greatly and my lips were raw inside, but somehow I managed to get started and from the moment that I dipped the goggles under the water I was too fascinated to notice my discomforts. The water here had a totally new clarity so that Chapters One and Two were blurred and muddy preludes to a crystalline brilliance which I had never believed possible. It was as if the water were air – so bright was it. This was particularly true in the shallow water where the coral began, about 20 yards off shore.

The next thing I noticed was that the fish were all quite different from those I had seen at Double Island – perhaps not all, but certainly most of them. Almost at once I saw the first blue fish. There were two kinds – one about 5 inches long, in which the male had an orange yellow tail and the female was all blue. They lived in pairs and their blue was of such brilliance that it appeared iridescent like a butterfly wing. The other kind was smaller with no orange tail and of a slightly different shape, though the blue was no less brilliant. These lived in shoals and were not, I think, the young of the larger species. When I saw the first shoal of a dozen or so I found myself exclaiming with astonishment out loud – it was a sort of shout of joy at seeing something so incredibly beautiful.

Charles swam with me for a good deal of the time. We found some huge clams (*Tridacna gigas*) which were not much less than 3 feet long. Then there were the deep ultramarine blue starfish. Charles dropped one into a big clam which closed convulsively, but without pinching the starfish, so that he later retrieved it. In deeper water the numbers of fish were greater but there was still nothing to compare with the two blue fish. One of the small, disc-shaped fish had bold black and white vertical bands and a long white streamer from its dorsal fin. And so gradually we swam out until we came to the launch. I had been swimming in a shirt so as not to get sunburned too much. I was now fairly tired, as we had been swimming for about an hour. We had seen nothing big in the way of fish but oh so many different small ones, such profusion, such variety.

My mouth was very painful so I decided to climb out on a rope which was looped over the side. In due course we were under way heading for Green Island. This is quite a different kind of island. It is much larger than Michaelmas Cay – about 32 acres – is covered with quite tall trees, and has a jetty at the end of which is the Underwater Observatory. The above water part of the observatory is as trippery as one would expect to meet in any tourist resort. But the idea is superb.

We were let in free and went through the turnstile and down the stairway into the underwater chamber which had been brought to the spot.

It was a large room with round portholes in all the walls and Phil and I began to watch the fish. Eye level was about 10 feet beneath the surface. Although the effect was very much like an aquarium the 'tanks' were in fact the open ocean. Some of the corals had been brought to augment the existing reef from somewhere round Green Island, but there was nothing to retain the fish, which were free to come and go as they pleased.

Here the profusion of fish and of species was once more amazing. There were large parrot fish, and coral trout, shoals of striped and spotted fish with huge eyes which lay hiding in the fronds of coral, chaetodonts with three or four different patterns and bright yellow snouts, busy little wrasse, large shoals of tiny fish looking just like the hatchet fish of freshwater tropical aquaria, which hung motionless opposite two of the portholes, the whole shoal swaying gently forwards and backwards with the movement of the waves. There were shoals of blue fish, not so bright as the ones on Michaelmas Cay, with sharply forked yellow tails, and in the background shoals of larger fish which looked like silvery pike and were all the same length – about 18 inches. And outside some of the portholes were huge sea anemones with their attendant anemone fish, the most brilliantly coloured of all – there were two species of anemone fish – one was the common one – orange with pale blue stripes edged in black, the other was dark chocolate with yellow fins and tail and a single brilliant opalescent blue stripe behind the head.

We did not have very long at the observatory but so impressed were we with Green Island and its potentialities that we decided to come back to it on the morrow.

Tuesday 25 December. Christmas Day. We set off again, this time in the tourist boat, *Mingela*, for Green Island. These two visits to the observatory constitute the Appendix to our underwater adventure.

Lloyd Grigg, the proprietor, pointed out that the chocolate and blue species had a single tiny baby already living among the anemone tentacles. This was intermediate in pattern, probably showing a basic affinity with the commoner orange and pale blue form. A third fish, intermediate in size with more orange on the body, was, Grigg told me, last year's young one. He said they laid twenty or more eggs but that only one young was reared. He even thought the parents ate the others.

He knew many of the fish by their scientific names but couldn't explain the number of differently patterned chaetodonts. The number of fish species in sight seemed even more than yesterday. I set out to count them and by going from window to window, trying not to count the same one twice, I found no less than fifty-two different fish species. These included many which I had not noticed the day before, including a sting ray, black above and white below, lazily flapping his wings with his long whip tail trailing behind him. He was far out over the sand

and periodically cruised into view. Then there were the razor fish which hung head downwards in a group of two or three.

After lunch we set off to the beach on the north side of the island. This fourth and last chapter of our underwater adventure consisted of about two and a half hours in wonderland. Phil was doing it all for the first time and she had some trouble with her mask and snorkel, also her flippers gave her cramp. About 10 yards out in 4½ feet of water there was a species of *Zostera* growing which had quite a few fish, but the coral did not begin for another 10 or 15 yards in a depth of about 5 feet, which was just too deep for Phil to stand and adjust her mask. At first I stayed with her, but later I realised that she would prefer to work it out for herself so I set off to join Charles on a protracted cruise of the nearby coral.

Just as each of the preceding chapters had outshone the last so now this cruise was far and away the best of all. Once again most of the fish were new: different from those at Michaelmas Cay, and not even identical with the ones at the observatory, which was about 200 yards away – or at least, not all of them. One of the first fish I saw, and one of the most colourful and amusing, was one which looked rather like a pig, with prominent blue stripes between the eyes and across the tip of the snout, and yellow patches elsewhere with a very characteristic black pattern. These ran up to about 10 inches long and were quite charming in their manner.

Cruising very close to the beach was a large fish, perhaps 2 feet long with a very blunt head. Perhaps this was the fish sometimes known as a dolphin. Also near the shore I came upon some grey mullet which looked just like the ones one might see in England. On the way out to the coral there were some sand gobies, more or less white with black spotting. There was a similarly marked blenny and another all black blenny, which lived in the coral and sat with tail bent in the typical blenny attitude, and yet it was surprisingly flattened and very conspicuous. Superficially it was very like another black, plate-shaped fish and I wondered if this was merely convergence or perhaps a case of mimicry.

One of the loveliest sights was the shoals of blue fish – not the very bright ones I had seen at Michaelmas Cay, but a soft blue fish which was in shoals always near a coral the tips of which were just the same colour. Presently Charles came and said he had seen what he thought was a young hammerhead shark, which had 'scared the daylights out of him' but if two of us went back . . . but alas he had gone. Charles thought he must have been about 5 feet long. I swam on out and came upon a large ray flapping lazily along. He was pale brown and spotted and had round-ended 'wings'. Later Charles and I found another hiding under some coral near the edge of the reef.

I went back to see how Phil was getting on. She had discarded her flippers because they gave her cramp, but she had mastered the mask and snorkel and had reached an area of coral where she could see quite a lot. I went off again, farther out on the main reef, and found first a new kind of anemone fish, brownish with three pale blue vertical stripes. This was to some extent intermediate between the two we had seen from the observatory. Whether these are different species or polymorphic phases I do not yet know. Next I came upon a pair of razor fish floating with head vertically downward. I gave chase to see whether I could scare them into any other swimming position, but I couldn't. They kept ahead of me still vertical, apparently by the use of a well developed anal fin.

Then I went out to the edge of the coral. Crawling slowly up and down over the sandy ripples of the sea's floor was a hermit crab in a tall shell about 5 inches long. Near the edge of the coral there were shoals of parrot fish – one or two rainbow-coloured males among a dozen drabber females. They flapped their wings up and down just like bright parrots. Among them at the edge of the coral were some even larger, deeper fish with little horns above their eyes and snouts like pigs.

On the way back we met pipe fish, small and spotted, a

large blenny with a pearly, lustrous eye, and finally a moray eel – white with black spots, which scuttled away under some soft coral and weed and later went farther still when I fanned him with my flippers. There was also a little fish with pouting mouth which fed along the coral in a vertical position, head downwards. He was not so well adapted for his attitude as the razor fish, and when swimming from coral to coral he came down to about 45°.

When we got back to the shore we had been diving for two and a half hours. I had finally adjusted my snorkel in such a way that the whole mouthpiece was inside my mouth and my teeth gripped the breathing tube. It was the only possible way with my lips so swollen and lacerated as they were. As I walked up the beach I felt quite groggy and unbalanced, but it soon wore off. While dressing, Phil pointed out the cicadas in the tree above our heads. There were dozens – on every twig they sat squirting out a little water from time to time.

We only just caught the boat and had a pleasant homeward journey, upon which we saw a flying fish scuttering over the surface. It had been a wonderful and memorable Christmas Day.

From *Animals* magazine

9

Sightseeing in East Africa's Game Parks

Wednesday 25 January 1961. At Murchison Falls National Park in Uganda. A day with John Savidge at Buligi – and *what* a day! Elephants, rhinos, carmine bee-eaters, and never a dull moment. We set off in the deluxe Land-Rover, made for the Queen Mother's visit, on the road to Buligi and the shores of Lake Albert. The first stop was to watch a very pretty, almost white hawk in a tree top. It had red eyes and eyelids and a short tail which wagged up and down incessantly and comically. The next stop was for a side-necked terrapin seen crossing the road in a very dry area. We took him on with us to the next water we came to, a very small pond. There were plenty of oribi and kob, and in some areas Jackson's hartebeest in larger numbers than we remembered from our previous visit; though this time visibility was extended because the grass fires were more advanced than last time.

So down to the West Nile, or Albert Nile, to the place where the Queen Mother had her elevenses, and where there was a profusion of dragonflies with black velvet bars across their wings. They were concentrated in the shade under a tree in which we saw the display of a striped kingfisher. We then came to the first tree full of carmine bee-eaters. Later we came upon more and yet more. At one point they were hawking round a group of elephants. We lunched under a bush near the first tree, and subsequently photographed 'orange tip' butterflies of two probably unrelated species. The larger one had deep scarlet tips to its wings while one of the smaller species had been caught and was being eaten by a jumping spider.

At Buligi, where the road bends south, we saw some reed-buck with black noses and short, forward-curving horns, and a bush-buck with a conspicuous pattern of white spots. At a ranger post we heard of two rhinoceros near the middle road along the peninsula and went back till we found a mother and large calf under a tree about 80 yards away.

We then went back to Buligi and on round, seeing a martial eagle with yellow eyes, an osprey, four kinds of plovers, and yet more carmine bee-eaters. We saw one of these eat a big black wasp with yellow antennae, beating the insect sideways against the branch it was sitting on. Others ate dragonflies and a butterfly. There seemed to be two or more species of yellow wagtails: a typical blue-headed one; one with longer legs, a larger bill and an all black head; and immatures which might have been yet other forms. We circled a small sized bustard and later, along the edge of the bay or lagoon, by which hundreds of sand martins were migrating, we saw the beautiful cinnamon-chested bee-eater, a small green one with a blue bar across the upper breast and reddish below it.

I have not described more than 10 per cent of the bird species seen, and made no mention of the waterbuck and the warthogs. It was a memorable day.

Thursday 26 January. John Savidge took us up the river in the outboard motor boat used for catching crocodile poachers. Particularly interesting were a flock of 325 skimmers, with bright orange bills; in the heat many of them were sitting on their knees. At one place we landed where crocs had been nesting. There were no crocs, but a monitor lizard about 2½–3 feet long. We continued up to the gorge and the incomparable falls, where there was 'the largest croc in the river', on a beach of his own on the left, and a rock with pratincoles on it.

On the rather showery return trip we saw four rhinoceros, and got quite near a troop of baboons. While we were away one of our gravid chamaeleons gave birth to eleven tiny young, all beige coloured and enchanting.

After dinner we went off on a frog hunt to a swamp on which pygmy geese had recently been seen. On the way we saw hare, rabbit, oribi, buffalo, and several species of nightjars. The 'frogging' was fairly successful; four species of frogs and a toad. But the great feature of the evening was a standard-wing nightjar on the way back which allowed us to walk up to within 6 feet, when it was dazzled by headlights and torches. I thought that the huge appendages must be analogous to the tribal development on the back legs of certain hemipterous bugs, and be designed to encourage an enemy to pounce at the wrong target; it seems possible that they can be moved independently of the bird itself. However, according to John Williams, the distinguished East African ornithologist, they are purely for display. There were also long-tailed nightjars and at least one other kind.

Friday 27 January. More procreation! An egg capsule of a mantis, which we had brought in from Buligi, had hatched and the inside of the window and back of the curtain were smothered with several hundred minute mantis nymphs. The chamaeleons, both large and small, paid no attention to the little mantids. One baby chamaeleon tried to eat a tiny fly and missed. We fed the larger ones; one ate thirteen house-flies in a row.

At about 10.30 we set off up the river in the outboard motor boat. On the skimmers' bar was a spoonbill with crimson face and pink legs. Later we got very close to a very large croc, which sat with open mouth, to a range of about 15 feet.

As we came into the gorge we saw large soft-shelled turtles on a rock. Nearby on a beach were four monitors. The two largest were digging and occasionally fighting. Eventually they came to a nest of soft-shelled turtle's eggs and began to eat them. We positioned the boat so that eventually we were only

10 feet from the largest monitor, which had now seen off its rival (which had wandered away causing a thicknee to give a distraction display in front of it). The remaining monitor was not less than 4 feet long. He was so hungry that he overcame his timidity, moving away as we approached, but always moving back, with his hugely long, blue, probing tongue. At the beginning he spilt some yolk and ate a great deal of sand with what he picked up, but later he got better at swallowing the eggs more or less whole, and we watched him eat about seven. Finally, hunger no longer overcame caution and he went off.

Just below the falls was a treeful of black and white colobus monkeys but the current was too fierce to dare to go close. Again there were pratincoles on the rock in mid-stream, and some near the mooring place for the boat.

On the way back a predominantly white plover flew over and landed on a spit upstream. We turned back to examine it and I made a quick sketch of it on the bottom of a fishing reel box. As soon as we got home I painted it. It turned out to be the whiteheaded plover, one of the ones alleged to pick crocodiles' teeth; it had not previously been recorded in the Murchison or Queen Elizabeth Parks.

Sunday 29 January. Back at Entebbe. A morning of thunderstorms and no flies for the baby chamaeleons. By 10 a.m. it had cleared and John Blower, at that time the Chief Game Warden, took us down to the Game Department by the edge of the lake where they have made the beginnings of a nice little zoo. Without doubt the nicest animal in it was a baby bush-buck only a few days old. There were also some very tame situtunga fawns and some buffalo calves, one of them complete with two ox-peckers. We went in with the delightful little elephant's child which had a taste for mangoes, of which we had a bucketful, but did not like strangers much, and finally charged Phil and pushed her back against the barrier, which was not a little alarming, though she took no harm from it. We caught up some small crocs, one of which bit me as I was trying to change two round. It barely drew blood on the ball of my thumb, but it was quite painful.

John Blower took us up to the Virus Research Institute to lunch with Dr David Gillett and his Polish wife Irena. He has done incredible work with mosquitoes. By using a method of blood transfusion in the mosquito that he has perfected, he has discovered that hormones released in the head after a blood meal, in turn release others in the thorax which bring on gonad development. If the head is removed and the blood of a mosquito which has fed is injected between four and eight hours later into a headless mosquito which has not fed, ovarian development results. Headless mosquitoes can be persuaded to copulate more easily than if they have their heads, because of the absence of inhibitory stimuli such as come from their sight and antennae.

Gillett had set a light trap the night before and had caught four good sphingids and one battered saturnid. One of the hawk moths was the vast blackish *Nephele equivalens*, nearly 4 inches long when at rest, and another was a very beautiful newly hatched specimen of the dark green *Nephele comma* with a white comma mark on the wings.

We spent a most delightful afternoon with the Gilletts. In the garden were a pair of double toothed barbets, and we met for the first time the ants which all beat dry leaves with their abdomens in unison, giving the effect of a snake going through the underbrush. It was a beautiful adaptation to protect their nest. We also opened a small cocoon, hanging from a branch 6 feet up and found it was a nest of very minute red ants.

Monday 30 January. Mervyn Cowie met us at Kenya's Nairobi airport and we saw our first giraffes on the way into the town. There were also zebras and wildebeest which had strayed from the Nairobi Park because of the drought which affected the food supply.

About half past four Mervyn drove us out to his home, 11 miles from Nairobi, through the Park. The animals were delightfully tame. There were giraffes, zebras, and several antelopes that were new to us: impala, kongoni (Coke's hartebeest), Grant's gazelle, Thompson's gazelle, and wildebeest. After tea Mervyn took me out in the Land-Rover, and we saw all of these again plus eland and a troop of sixty-two baboons. The low afternoon light was very beautiful, and the animals were delightfully close. It was a marvellous show.

Tuesday 31 January. Mervyn took us into the Park after early breakfast and we found a pride of lions, consisting of two adult lions, known as 'the spivs', two young lions, and six young lionesses. On the way we saw a pair of silver-backed jackals.

At the Coryndon Museum I was taken round by Dr Leakey, director and archaeologist, seeing good dioramas, Joy Adamson's paintings in water colour of flowers and portraits of native tribes and their tribal dress, a new technique of casting snakes on branches which produced remarkably life-like models, bird, fish and insect galleries, and reproductions on artificial cave walls of wonderful cave paintings. Previously he had shown me the new 'Snake Park', in the charge of his son, with live snakes, chamaeleons, and frogs. They told me there were thirty-nine kinds of chamaeleons in Kenya alone and I saw two half-grown specimens of the largest which grows to 3 feet long, but these were only about 2 feet long, though impressive.

Dr Leakey, whose discoveries in archaeology and palaeontology are of first importance, took me up to the Bird Room to meet John Williams, who was vastly impressive and very charming. He began by presenting me with a tule goose skin collected by Jim Moffitt in 1940 in Solano County, California, which he said we ought to have. It was a delightful gesture. He also gave me skins of carmine and blue-cheeked bee-eaters.

Dr Leakey showed me 'the most marvellous example of mimicry in the museum': Hemipterid bugs that were pretending to be flowers. They existed as a colony of about twenty to thirty, were bright yellow, and clustered at the top of a herbaceous plant which bent over with their weight and hung down. But most remarkable of all, every colony contained from one to five individuals which were green instead of yellow and these green ones were always at the end of the plant representing buds. If the colony was disturbed the green ones would be in the 'bud' positions within half-an-hour. How do they know? There is a similar white species, but it has no buds.

Wednesday 1 February. We went with Mervyn in the Land-Rover to Tsavo Park West; we motored 25 miles to Mzima Springs through quite pretty country but entirely devoid of mammalian life. It was extraordinary that there were no great herds of antelopes, as the grass looked lush enough. It seems that there were too few available water holes, and bore holes and water troughs are being installed to try to bring back the herds which were poached out many years ago.

Mzima Springs, with tame vervet monkeys, were very lush and green with clear water gushing out of the lava rock. There were raffia palms, papyrus, and great blue chub-like fishes (*Barbus*). Farther down, in a wide pool we watched half a dozen hippos in a clear pool from an observation tower with a covered approach. A cormorant was fishing under water – at first I thought it was a fin-foot. Farther down still was a cleverly designed submerged tank with windows through which a shoal of about a hundred *Barbus* could be seen at very close quarters; the largest was about 5 pounds. There were also a few large 'sucker fish', some tiny *Tilapia*, and some small loach-like Eleotrids. We dug out and photographed some ant-lions at the bottom of their conical traps, and also found a wingless wasp mimicking an ant.

After dark the return road was full of nightjars – one of which flew into the car and killed itself, so we brought it back and drew it. Its eggs and nest are undescribed; evidently it is very local. A hare (which was smaller than the European hare) ran in front of the car. Later near the Warden's house, where we were staying, the headlights showed an elephant and a herd of forty impala.

Thursday 2 February. We left for Tsavo Park East, where we were met by Peter Jenkins, Assistant Warden. We drove about 20 miles through the Park to a camp specially set up for us on an island in the river. On the way we saw several groups of elephants, covered in the local red soil and looking almost chestnut-coloured. I saw a single eland, and we passed many impala and waterbuck, a lioness and two cubs, an Egyptian goose with three full grown young, and lilac-breasted and European rollers. The Athi River was dry but the Tsavo, fed by Mzima Springs, was still running and there were certainly many more animals concentrated by it than we had seen in Tsavo West.

The camp was in a most romantic setting under Doum palms and we crossed a sandy river bed to the island. Beyond was a cleft in the escarpment of the Yatta, the longest lava flow in the world. Through this pass was a well worn track from the waterless plain beyond to the river, and the site was selected for this reason.

At camp was the warden of Tsavo East, David Sheldrick, his wife Daphne, and their little girl Jill, aged six. The camp was set up in luxurious style and we had iced drinks followed by a slap-up lunch. A character of special appeal was the Sheldricks' young banded mongoose, which was absolutely and deliciously tame, whistled almost perpetually, and would break an egg by throwing it backwards between its hind legs against a box.

There was a cry that elephants were approaching. They were still up in the 'pass' but coming fast towards the river, evidently intent on a drink. We took up a position on the bank in the shade of a bush, seven of us, including Jill, and waited for the elephants to arrive. Eventually they came to the far bank of the river just below us and only 50 yards away. At once they began to drink and splash water over themselves, occasionally lifting their trunks to sniff the wind, on which perhaps some evidence of our camp and ourselves may have eddied out towards them. For half an hour we watched the nine of them.

One of the elephants crossed the river to our island, and the rest followed. Evidently they contemplated spending the afternoon in the shade of the Doum palms. They turned a little towards us along the bank. We began to withdraw, and the oldest cow saw us. She came determinedly up the bank towards us. We all began to run, and David Sheldrick and Peter Jenkins turned and pelted them with Doum nuts. This had little effect and the elephants headed on towards the tents. About 20 yards short of them they turned and skirted round one end of the camp as David Sheldrick started up a Land-Rover and rushed them, to speed them on their way. I do not really know how dangerous the situation was, if at all, but it was certainly very exciting for a couple of minutes.

David Sheldrick dug down a scorpion hole, which was completed by a ranger, and found two scorpions about 3 to 4 feet from the entrance and round a right angle corner. The second was nearly twice as big as the first and spanned the middle part of a soup plate.

Later we crossed with the Land-Rover to the east side of the river and went up the shore track, seeing impala, waterbuck, the lioness with her two cubs – all very thin – a fleeting glimpse of a rhino with calf, a herd of buffaloes, some hammerkops, woolly-necked storks, marabous, a fleeting glimpse of two lesser kudu females, and more distant elephants. But the most interesting feature of the drive was the variety of small birds: doves, including a pretty pintailed one with red-brown wings; a sand grouse with black wings; a new green bee-eater (said to be the Madagascar bee-eater), three species of small hornbills, red bill, white bill, and parti-coloured bill with a black-billed female; yellow-breasted glossy starlings with long tails; two rollers; drongos; shrikes; and flycatchers.

We were back in camp before dusk, and after a drink ourselves, we went 50 yards to the edge of the river to see the animals which had come in to drink at the river's edge. There were eighteen elephants stretched across the silver track of the full moon which had just risen above the Yatta escarpment, and standing by the waterside was a rhino. All these animals were within 60–70 yards, silhouetted against the moonlight. It was inexpressibly lovely and romantic.

During dinner we kept looking with glasses to see the elephants arriving and departing, and after it we returned to the river bank, to find three rhinos and rather fewer elephants. Soon after we had turned in they told us that there were at one time thirty-six elephants, but by 1.30 when I got up to have a look there were only eight: four departing and four arriving. The sucking noises of elephants drinking are very impressive. At a quarter to six, as the first light of dawn dimly appeared, I watched four more elephants at the water's edge. For these moonlit views of the great pachyderms alone this camp was more than worth while.

Monday 13 February was a day of hard work in Nairobi – a meeting with John Owen, Director, National Parks of Tanganyika, and then a meeting with the Trustees of the Kenya Parks.

Next, a broadcast interview to the Forces, in which I selected *A Bar at the Piccolo Marina* as my piece of music. I was not allowed anything serious, but afterwards they played me the last three minutes of *Rosenkavalier* which I would have selected. Having been starved of music for several weeks, I was greatly moved by this glorious passage, and could not see to do the little drawings in the autograph books which had been put in front of me.

After lunch I made a short introduction to a bird film to be sponsored by East African Airways and to be shot by John Pearson. He had brought a tame kestrel, a Woodford's owl, and a duiker, though we used only the kestrel in the film. A studio had been built in a hangar at the airport. It was very hot under the lights and I had to be scriptwriter, director, continuity man, the lot!

When we finally got back to Mervyn Cowie's home, Phil, who had had a quiet day, had caught a beautiful three-horned Jackson's chamaeleon. We took photos of it instead of going out into the Nairobi Park with the warden. It was really a lovely chamaeleon and Phil had found it in the garden by Cecil Webb's method – just looking in a bush!

Tuesday 14 February. Mervyn drove us to Nakuru in his big Buick, with his wife, Val, and John Williams who told us about the birds on the way. We were rather late arriving but nevertheless they took us down to the edge of the lake where the ceremony of opening the National Park was to be held in the evening. There were hundreds of lesser flamingos all over the shallow soda lake (which is now far below its normal level). The strong north-east wind was whipping up clouds of white soda and, as we stood beside the platform being set up for the ceremony, it was like a smoke screen in a war film. The dust was pungent and horrible. We lunched at the Stag's Head Hotel. Its proprietor, Norman Jarman, is one of the prime movers in establishing the National Park and was in charge of all the arrangements. Also at lunch were Jack Hilton, Mervyn's second in command at the Parks Office, and his very charming wife, and David Roberts who catches flamingos and crocodiles for zoos and for Armand and Michaela Denis to film. He had caught the two lesser flamingos which are to be presented to us but, for reasons of climate and import licences, they will not be sent till April. We are hoping that we can get some more sent at the same time to make a small flock.

After lunch Phil and I went off to a school to give a lecture with slides – a useful opportunity to get to know the order of the slides.

Then we went back to the hotel for tea and down to the

lakeside for the opening ceremony of the Lake Nakuru National Park. The wind had eased and the soda dust was not too bad. The Governor – Sir Patrick Renison – arrived with the Mayor of Nakuru and the Chairman of the County Council. We all mounted the rostrum and the speeches began. I did mine without notes, and tried to make it forceful. It seemed to go over fairly well, and Mervyn did a witty vote of thanks.

Then I was invited to join His Excellency in his brand new Rolls for a drive down the lake side. Mervyn followed with Phil and Val. The African driver did not dare to go very close to the lake edge, so when we turned it was agreed that Mervyn should lead. Halfway back he almost got stuck and we swung away from the shore only just in time.

We had passed many thousands of lesser flamingos scattered rather loosely along the shore, with occasional ruffs and little stints. But, apart from the great number and the beauty of their shape and colouring, there was no greater interest in one stretch of shore line than any other. The soda dust which seeped into the car was stifling. As we came to the place where we should turn inland, Mervyn overshot the route and in a moment he was inextricably bogged. I shouted to the driver of the Rolls-Royce to stop and we pulled up only just in time.

The police arrived with a Land-Rover and 50 yards of rope which parted every time it took the strain. Eventually, with lots of people pushing, Mervyn's great Buick was extricated. But, as a result, we were almost late for my lecture in the Town Hall.

Wednesday 15 February. A Flamingo Day. Before sunrise we were driven out by Norman Jarman and Tony Dando to a spring at the edge of the lake. Coming down towards it we passed a herd of nearly a hundred impala, some guinea fowl, and a single waterbuck. In the spring, surrounded by thick vegetation, we were shown six of the twelve hippo living on Lake Nakuru. They are in some danger as two have recently been shot, and the spring is just outside the new National Park.

We waded nearly up to the top of borrowed wellingtons through soft mud and a bed of tall reed mace till we emerged at the edge into a hessian 'hide' of primitive design and construction. The sun had not yet topped the hill behind us. The far shore of the lake was already sunlit but the uncountable masses of flamingos which stretched from the far distance to within twenty yards of us, glowed pink in the blue shadows.

On our extreme right, where the stream debouched from the reeds, was a group of half a dozen Hottentot teal, but they had seen us arriving and they were soon away. There were also marabou storks which flapped away and cleared the immediate area of flamingos, though before long they were walking back towards the hide.

As the sun rose behind the hill the colour became more and more brilliant. Streams of flamingos came in from far out on the lake, a triangular patch of bright scarlet under their wings lit by the low sunlight. There can be no more remarkable ornithological spectacle in the world.

We stayed in the hide for two and a half hours and Phil took many pictures. In the foreground there were sometimes blackwinged stilts, often ruffs, and most of the time, little stints. There was also a female shoveller which sometimes swam almost amongst the flamingos' legs.

After breakfast we set off for Lake Elmenteita with Mr and Mrs Jarman. The flamingos there were much more scattered but there must have been a hundred thousand or more. A small proportion of them were greater flamingos, which are much wilder and more spry than the lessers. Their calls are much more goose-like, reminding me of greylags, while some of the lesser flamingo calls are very like those of pinkfeet. (A baby lesser at Lake Nakuru, still in its grey plumage and much smaller than the adults, made a chirruping call which was strongly reminiscent of a gosling's call. I am sure they are just long-legged geese.) Our efforts to take pictures of greater flamingos were not very successful. New birds here were great quantities of avocets (many hundreds in one bay), scattered Cape teal, and several hundred shovellers.

In the evening we went back to the hide by the spring and the flamingos were perhaps even more impressive against the light. By the end of the day Phil had taken three films of colour and two of black and white and the female shoveller had collected a chum – a female garganey.

Thursday 16 February. Norman Jarman took us to the famous Menengai Crater from the rim of which was a magnificent view. We saw a falcon of some kind ridge-soaring, and a tawny eagle thermal-soaring. Then to the Nakuru Municipal Dump to photograph marabous, which are very impressive in the air. The wing span must be 9 or 10 feet. There were 150 or more on the rubbish tip.

Immediately after lunch we set off with the Jarmans for Nairobi and saw ground hornbills and giraffes on the way.

Friday 17 February. After lunch John Pearson took us out to his home to photograph some of his tame animals. The duiker, Tikki, was the most appealing. He has had it for four years and it is completely and charmingly tame and exquisitely beautiful. She is a female and about the size of a Labrador retriever. The Pearsons also have a long-coated golden retriever which often shares a kennel with Tikki.

Inside some bamboo aviaries Phil took pictures of an eagle owl, a Woodford's owl (called Woodford and borrowed from some friends), and finally, in the open, she photographed Charlie, the adult female African kestrel, in big close up against the sky while it sat on John's gloved fist. Tikki, Charlie, and Woodford had been at the improvised film studio in the hangar at the airfield when I filmed the introduction to John's as yet unshot film of African birds a few days ago.

Saturday 18 February. We spent the morning in the Nairobi National Park. The concentrations of animals were astonishing. In two places we saw seven species in company – once at a salt lick, once at a dam. On the first occasion there must have been between two and three hundred animals: zebra, wildebeest, kongoni, impala, Grant's gazelle, waterbuck, and ostrich. At the dam were zebra, wildebeest, kongoni, Grant's, Tommies, giraffe, and warthog.

A ram impala with fifteen ewes was especially co-operative in photography as he and his wives stood in the shade of a tree. Mervyn said he had an exceptionally good head. They must be the most beautiful of all the antelopes.

Across a valley, we saw an ostrich running, pursued by a small four-legged creature. We could not think what it could be; perhaps a hyena, or a jackal. We converged on its course and found it to be a day-old wildebeest which had mistaken the bird for mamma. The ostrich, embarrassed, had trotted away and the tragic little chase had begun. By driving the Land-Rover between them (they were only 10 yards apart) we managed to separate them, but there seemed no way to get the little animal to go back to where it came from, and anyway its mother may well have been dead or dying. The Park was littered with corpses due to the drought and lack of food. It is already serious and will be disastrous if no rain comes for another month which may well be what happens.

Tony Parkinson, partner of John Seago in his animal business, came to the Cowies' to take us to lunch with John. Tony's wife, Thelma, was there, and young Julian Tong, recently arrived from England, who has joined for the catching operations. John Williams called in after lunch and told me that the grey falcon we had seen at Nakuru was probably a lanner and that the harrier eagle was probably the black-breasted.

In the Seago-Parkinson Land-Rover, which had known better suspension, we set off – six of us – on an excursion to the Ngong hills where the game has been pushed back by Kikuyu settlement. Eventually we found herds of buffalo, eland, and waterbuck high on the slopes above us. Later, and nearer to the road, were zebras, kongoni, Grant's, eland, a giraffe, and a reedbuck – the first we had seen since Murchison Falls Park.

We motored across country to the highest point of the Magadi Road – more kongoni and zebras – and a superb view across the Rift Valley with many giraffes far below us. On the way back through the Masai reserve there were big herds of wildebeest, a few eland, quantities of zebras, an adorable lone baby Tommy, and a herd of about fifty half-grown ostriches accompanied by two adult females. One of the young was much smaller than any of the others.

So back to the Cowie home at dusk – much shaken up!

Sunday 19 February. I went with Bill Shepherd to Lanet Airfield, Nakuru, for soaring. I did not manage to soar, though I think I should have done in a more familiar glider. Twice I was in small rather broken thermals and had to give it up because I had no idea of the Grunau Baby's performance or of its spinning characteristics.

Monday 20 February. In a two-car convoy with the entire Cowie family and John Williams, we went to Amboseli – with lunch at Namanga, 100 miles from Nairobi.

John saw thirty-one species of birds before lunch. I saw only twenty-four of them.

After lunch we passed through gerenuk country. These extraordinarily elongated gazelles are very difficult to see, and, as we had missed them at Tsavo, we were very excited to see the first one. Altogether we saw thirteen – one party of five and another of three with a large buck. The rest were single females. We also saw a steinbok, small and reddish brown with big ears.

A rhino greeted us as soon as we had crossed the dried up Lake Amboseli. Some new springs, emerging as a result of seismic activity only a few years ago, have provided an oasis between the old lake and Kilimanjaro which is over the border in Tanganyika.

The camp consists of a number of bandas (small huts) and the warden's house in which we are staying. The warden is 'Tabs' Taberer, and his assistant is young Charles Moore. Tabs has a tame bat-eared fox of great charm.

Soon after our arrival we set out again with Tabs and went to a causeway across an artificial river, canalised for 7 miles out from the springs. Above the causeway was a great profusion of water birds, including four species of ducks – redbilled pintail, garganey, African pochard, and white-backed. There were also squacco herons, egrets, a dwarf cormorant, lily trotters, black crakes (which are tiny moorhens with yellow shields and red legs), African moorhen (much like ours), and a considerable rarity – the Kaffir rail – which looks like a large water rail, the size of a moorhen.

The long-toed lapwing, which is also very rare according to John, is adapted for walking about on floating vegetation, and the pair we found were doing just that. They were very tame and Phil got some pictures of them which John asserted were the first still photographs ever taken of this species, although Tabs had taken some 8mm movies of it.

During dinner there were ant-lions coming in to the light, and a most marvellous little moth which imitates a couple of thorns. The upper wings wrinkle up so as to become thorn-like and the lower wings 'fair off' the insect on to the twig. Its name is *Sesquialtera ridicula*. We kept the moth and a couple of ant-lions for photography the following morning and did them before, and immediately after, breakfast on . . .

Tuesday 21 February. In the morning Kilimanjaro was cloud-free and the crater was as magnificent and snow capped as we expected. It does not remain in sight for long at this time of year, and does not always appear even in the morning. We invented a new greeting for Amboseli: 'Top of the mountain to you!'

We found Gertie, the famous rhino cow whose especially long horn has now been broken. She was with a two-year-old calf and one four years old and the three of them were all lying down. The younger calf laid its head charmingly on its mother's. They were so tame that we photographed them from the Land-Rover at 15 to 20 feet. Eventually they got up and sauntered 50 yards away to start feeding. It was a better view than you could get in any zoo and infinitely more satisfying.

Then we drove down the side of an artificial canal dug in order to carry water out to the Masai cattle. There were Grant's and Tommies all the way along and quantities of birds: crowned and blacksmith plover, glossy ibis, redbilled pintail, and Hottentot teal. The redbills sat to about 15 yards. The Hottentots were substantially tamer. There were also stilts and wood sandpipers, and a pair of crowned cranes. The canal ends in a great cattle-trough and a lake of about 100 acres. Around the lake were Egyptian geese and marabous and in and around the shallow water were quantities of waders. A flock of stilts – perhaps thirty of them – looked predominantly black; a flock of fifty avocets looked predominantly white. The commonest waders were probably ruffs and reeves and wood sandpipers but we also saw three or four spotted redshanks, a little ringed plover, and some Kittlitz's plovers, a spoonbill, and a flock of garganey.

We returned round the lake and along the other side of the canal, seeing a concentration of several hundred vultures drinking and bathing and a pangani longclaw – rather like, but unrelated to, the American meadow lark – which John said was very rare. Phil took photos of it showing its yellow throat at a range of 7 feet. We met a male rhino which trotted off and eventually met unexpectedly, and rather aggressively, a second rhino, which turned out to be Pixie – a son of Gertie, born with no ears and half a tail. The two faced each other for a while and then the original one turned back and trotted away.

On a branch of yellow-barked fever tree (*Acacia xanthophilia*) we saw a Verraux's eagle owl, but he would not have us close enough to get photographs. The wildebeest herds had a great many young, only a few days old. They were pale fawn with dark heads and an almost human voice of great sadness.

In the afternoon Charles Moore joined us again and we went round the eastern swamp, starting with the 'Rhino Run' at the end of which we met two moderately tame rhinos. Later, a Hartlaub's bustard walked across in front of the Land-Rover with dignified and unhurried gait. John had to make it fly to be sure of the identification.

The zebras were barking and moving off to the south of us as we photographed some giraffes. As we went to investigate there was a great wave of plains game – zebras, wildebeest, Tommies – moving away eastwards and soon we found the cause – a lioness which appeared still to have hidden unweaned cubs. She sat out in the open, and when we stopped within 10 yards she yawned, got up and walked 10 yards farther, flopped down and yawned again.

Round the far side of the eastern swamp was terrible erosion and we came upon a herd of Masai cattle. Seven thousand are grazed in the Amboseli area. We found one too weak to stand, which had just been abandoned, and would obviously be killed by a lion during the night.

Parties of vervet monkeys were climbing a tree. The increasing swamp had a border of open water with many waders, including Temminck's stint, marsh, wood and green sandpipers, ruffs, a greenshank, and European and African snipe. A pair of little bee-eaters lived up to the reputation of their family for brilliance of colour and elegance of shape. Earlier we had seen two species of pratincoles (plovers pretending to be swallows) and a very charming little bird, the Masai double-banded courser. Kittlitz's plovers had almost fully fledged young.

In the dusk before dinner a rhino came to drink in the water hole in front of the camp. At dinner a marvellous but horrific spider ran across the floor. This was the termite-eating solpugid spider *Galeodes*. It was captured under a glass and gnashed its enormous, independently operating chelicerae. We kept it for photography on the morrow.

Wednesday 22 February. Mervyn and I were to have flown to a conference of Veterinary and Forestry Departments of all East African territories at Lake Manyara in Tanganyika, but rain there in the night had made the airstrip unserviceable and led to a cancellation of the whole plan. Instead we decided to concentrate on photographing small creatures in camp and to set off for Nairobi about 11 a.m.

Phil took pictures of the exquisite young bat-eared fox while I hovered in the background with the horrific spider, which in due course had to be run over my hands for scale. It was soft, apricot coloured, with long golden hairs scattered sparsely along its legs. Its eyes were too close together and its jaws looked highly dangerous. Yet somehow it was ever so slightly pathetic. I think it must have been a very large female. John Williams said he had never seen such a large one. He said the hairs broke off and caused a serious rash, much worse than any caterpillar. He said he wouldn't like it running over *his* hand – but I took no harm from it.

Phil photographed in quick succession superb glossy starlings, a clawed toad (*Xenopus*) which was caught in the swimming pool, the tiny *Lygodactylus* geckos which are diurnal, and a huge, fearsome, black horsefly. But by far the most exciting subject was a Lycaenid butterfly *Aphnaeus drucei* which was feeding on the flowers of a poinsettia in the garden. There were several, but the best female had small double swallow-tails on the hind wings. When the wings were folded these swallow-tails splayed out a little so as to look exactly like the insect's head, with two long, yellow tipped antennae and two shorter ones inside. The insect moved its hind wings slowly when at rest so as to simulate the movements of the head of an insect, in order, no doubt, to encourage a bird to grab in the wrong place. The life history of *Aphnaeus* is no less marvellous, for its larva is reared in an ants' nest and milked by the ants, which, in return, offer it their own larvae to eat.

Farewell to Amboseli. We saw a fine gerenuk ram and many birds on the way back to Nairobi, and a sloughed snake skin – tiger snake – at lunch time.

From *Animals* magazine

10

Conservation and Africa

The campaign for saving the world's wildlife is gathering momentum, and it is world-wide in its scope. Many animal species, evolved over hundreds of thousands of years, have become extinct at the hands of man (in the last hundred years nearly a hundred kinds of birds alone), and scores of others are on the very verge of extinction. The prospect of species-extinction is appalling in its irrevocability. Animals which have taken all those millennia to acquire their unique and exquisite adaptations to their environment, now stand in imminent danger of being wiped out in a few decades because people do not care enough to prevent it.

There are, of course, wider aspects of conservation than the extinction of species. There is the whole field of renewable resources, the whole realm of ecology – that intricate relationship between water and soil, plants and animals, and man himself. It may be that a proper understanding of the Ecology of Man holds the secret of his very survival. Some think that threatened mammals and birds and reptiles are getting a 'free ride' on the wider issue of man coming to terms with his environment. But that is surely a materialistic view.

Supposing we ask ourselves two age-old questions: What are we here for? What distinguishes man from other animals? Do we answer the first as most other animals would – 'To keep our bellies full and stay alive, at least until we have reproduced our species'? And do we answer the second by saying – 'These things we can do more efficiently than other animals by reason of our well-developed brains'? Surely we must have higher ultimate aspirations. Should not our answers rather be that we know the difference between good and evil, and aspire towards good, that we recognise beauty and can strive towards it, that we have a continuing curiosity and must seek always for truth? No one can doubt that our values are wrong if we think only of material welfare.

For conserving wildlife and wilderness there are three categories of reason: ethical, aesthetic, and economic, with the last one (at belly level) lagging far behind the other two.

The first argument arises from questions like this: 'Does man have the right to wipe out an animal species just because it is of no practical use to him, or is a nuisance to him? Are his belly-interests paramount? Is there an issue of right and wrong?'

'Supposing,' the argument goes on, 'that a man goes out and shoots the last Arabian oryx in the world, has he done something worse than if he had gone out and shot a Grant's gazelle or a wildebeest?' Many believe that taking animal life is morally wrong, but surely all would agree that killing the last individual of a species is *more* wrong.

The ethical argument can be completed with the quotation from King George VI which hangs over the main entrance to the Nairobi National Park:

> *The wildlife of today is not ours to dispose of as we please. We have it in trust, and must account for it to those who come after.*

The aesthetic case is a simple one. People enjoy animals; they find them beautiful and interesting, and often experience a re-creation of the spirit when they see them. To wipe them out is foolish and irresponsible because it deprives present and future people of a basic enjoyment. These arguments will not cut much ice with a man on a starvation diet. It takes a saint or a hero to put ethics (let alone aesthetics) before survival. To the large numbers of people in the world who are protein hungry, the economic arguments will inevitably be the strongest, even though they may be the least enlightened. But let those who are *not* hungry be quite clear in their minds that if conservation succeeds mainly on the economic case, man will once more, as so often in his history, be doing the right thing for the wrong reasons.

And what is the economic case? It rests on two main considerations – the tourist industry and food. Nothing is more certain than that travel will become cheaper and more popular, and that more and more people will flock to see the great wildlife spectacles of the world. Africa will assuredly draw thousands upon thousands to see the remnant – still a unique spectacle – of the great mammalian climax of the Pleistocene. A vast industry is waiting for development and it will bring riches to the countries which have taken pains to retain their wildlife.

Nowhere else in the world can one see – almost in a single eye-full – elephants, rhinos, giraffes, lions, zebras, warthogs, and perhaps a dozen kinds of antelopes.

If you put capital into, say, a copper mine, in due course you are left with a hole in the ground, but if you conserve your wildlife you will draw revenue from it in perpetuity; it is a 'renewable resource'. At the same time the lives of thousands of people will be enriched by seeing it.

For the most part this wildlife will be in National Parks, and I hope and believe there will be new developments in the presentation of it to the visitor. Nowadays people usually watch from a vehicle, and in those Parks where the animals see enough vehicles to become totally accustomed to them, the watcher may see them living their lives undisturbed; but in most places the vehicle causes some interruption of natural behaviour. In future I believe the more sophisticated tourist may require some sort of observation hut or hide into which he can creep (a bus load at a time if necessary) and watch creatures which have no knowledge of his presence. Such observation huts must be accessible by a covered approach so that the watchers can come and go without making any disturbance at all. This may require some ingenuity to devise, but should not be wholly insuperable. Such hides, coupled with improved exhibits at the Park gates or at the Lodges, will give the Parks an even greater meaning to the visitor than they have at present.

But what of the wildlife outside the world's National Parks? Can it, too, survive? This is where protein considerations come in. It has been shown conclusively that on certain 'marginal' types of land a greater weight of protein can be cropped from wild ungulates (that is, hoofed animals), without detriment to the breeding stocks, than would come off the same acreage under domestic animals. Furthermore the wild animals, having a variety of slightly different feeding habits and food preferences, do not easily damage the range in the way that domestic cattle do. Wherever this is true it is obviously wasteful and shortsighted to develop the land expensively for cattle.

Trypanosomiasis, carried by tsetse flies, is a serious disease in cattle, but the wild animals are immune to it, though the trypanosomes are in their blood. It cannot be sensible, on these vast areas of poor soil, to cut back the bush in order to get rid of the shade needed by the flies, and to slaughter the game in order to get rid of the reservoir of trypanosomes in their blood, when the protein derived from the cattle will be less than can be cropped from a healthy stock of antelopes. And in comparison, the risk of overgrazing will be substantially less under the wild species.

The conception of cropping wildlife is a sophisticated one which is not always easy to explain; yet it is likely to be necessary quite apart from the economic aspects. By comparison with former times a sadly small proportion of the world's surface is nowadays available to wild animals. If we practise conservation for any of the reasons we have discussed, and if we agree that 'conservation is for man', then one of the prime objects must be to maintain, for the enjoyment of man, the greatest variety of animals and plants in their natural association; and this postulates a degree of 'management' of the various populations and habitats so that the common ones do not swamp the rest.

At the present stage of man's moral and physical evolution, it may be that in order to save wildlife we must stress the tourist industry and the protein; but there is one other factor less immediately commercial than either – the question of national prestige. Nature Reserves and National Parks have become a status symbol. A modern state is not complete without its chain of them and its enlightened code of wildlife legislation. Governments will continue to want their countries to appear modern, complete, and enlightened. From this simple aspiration wildlife may yet derive additional security.

Recently Africa has been in the forefront of the conservation scene for a number of reasons, perhaps chiefly because of the sweeping political changes and because its wildlife is so striking and spectacular. Although only a remnant is left of the vast herds of mammals which formerly roamed the continent, yet in a few places the concentrations are still a staggering sight, both in numbers and variety. The birds, reptiles, amphibia, and fishes of Africa are also astonishing in their beauty and diversity – and so, too, are the invertebrate animals, especially the insects.

Africa has its share of endangered species. The quagga and the blue antelope, like the dodo, are gone for ever. There is a long list of others whose continued existence is by no means secure – a list which includes both species of rhino, the mountain gorilla, the okapi, the cheetah, the red lechwe, the dibatag, the red hartebeest, the bontebok, the white-tailed gnu, and the mountain zebra. For some of these rarities protection is already effective, indeed some only remain in semi-captive herds; but for others the situation is desperate.

In the conservation of threatened animals it is helpful to

recognise two categories: first, those species which are limited by the available area of habitat remaining to them, and which will fill out this habitat against the current drain on their numbers, and secondly those which are limited by a mortality rate that exceeds their reproduction rate. Thus on the one hand there is a threat caused by habitat loss and on the other a threat by direct depopulation, in which every additional animal that survives is significant. The conservation measures are very different in the two cases. By far the most animal species are primarily controlled by available habitat, and this even includes most of the very rare species in Africa. The other category (which includes both white and black rhinos) seems to be a late stage in the process of extinction, calling for urgent action.

And this is not only an African problem. All over the world the evolution of man and his technology has spelled disaster for living creatures – by reducing their available habitat and by killing too many of them. Yet this same evolution has evolved conservationists – from the enlightened philosophies of the world's gentle religions to the modern naturalist who takes delight in watching the living animal and the modern ecologist who brings science to bear on the proper relationship between man and his environment. Conservationists today are involved in a gigantic holding operation – a modern Noah's Ark to save what is left of the wildlife and wild places, until the tide of new thinking begins to flow all over the world. That it already flows in some parts of it, is clearly indicated by the immediate response in many countries to the campaign of the World Wildlife Fund, and by the widespread interest in the work of the International Union for the Conservation of Nature – particularly the technical help in conservation which it offers to the newly emerging African states.

As more and more people begin to think these issues are important, the threatened species will become increasingly secure. The time may not be far distant when all men will recognise the value of wildlife and wilderness to mankind, and will be agreed that these natural treasures must be preserved in perpetuity just as certainly as the great art treasures of the world. Then our tide will become a main stream in the evolution of man.

Introduction to *Animals in Africa*

11

To Record the Moment – or Enjoy It?

The sun was setting in the north as we flew back towards Flagstaff Island. It is noticeable as the season advances that the nights are much darker, and the sun now sets for an hour or so each midnight. This particular sunset shone bright red, out of a blue-grey cloud front, and was reflected in the multitudinous lakes. When we turned west along the shore, the scene was even more lovely. It was a sight to be seen and experienced once, and not forgotten. It is strange that the desire to keep, to preserve, to collect, to bottle the beautiful sights of the world is so strong. I suppose that it is good to wish to share their beauty with people who are not lucky enough to be there and see them, too. Or is this sententious? There is always painting – to add some of the character of the painter to the scene. There is photography; but here there is the danger that so much of the time is occupied with setting the camera that there is no time to experience the sight oneself. In the keenness to get a good result, one is tempted to say: 'I haven't time to enjoy this now, but I'll get my "potted" enjoyment along with my friends at home when we look at the photos together.' This must be disastrous, and yet I go on taking photos. And if the focus is wrong, or they are over-exposed, my excuse will be that I was trying to look at the scene, too, at the same time.

From *Wild Geese and Eskimos*

12

Anabel

I suppose the return of Anabel on the October morning was one of the half dozen most stirring events that I have ever experienced. The thrill was no less because it was half expected. I had never really doubted that she would come back, but when she did I was overwhelmed with joy and relief and wonder.

Anabel is a pink-footed goose. On 25 September 1936, she came first to the lighthouse which is my home. It was early in the season and the winter flocks of geese which live on the nearby marshes had not yet arrived. Soon after sunrise Anabel, led by that strange instinct which enables young birds to precede the old ones on the southward migration, came to the salt-marshes of the Wash. She heard some geese calling below her and swung in towards them. There were twenty-nine of them, sitting on the marsh quite near a strange round building; but they did not seem to be alarmed by it, so Anabel swept round low over them and called in answer and immediately a great babble arose. Anabel was very pleased to find others of her kind, and she circled round again. She did not realise that they were tame pinkfeet which had lived several years at the lighthouse. As she passed a little plantation some rooks came out and mobbed her; but she was too tired to care, so she curled back with set wings and settled near the pinioned geese on the marsh. They didn't greet her in a very friendly way, considering that she was one of their own kind. Some of the more pugnacious ganders chased and pecked her, but Anabel was too tired to run away and so she just crouched down and waited till they had satisfied their anger. Then she stood up again, preened herself, flapped her wings and walked after the other geese towards a fresh water pool, which is just above high tide mark on the side of the sea-wall.

On her way to the pool she walked past me, as I stood watching from the top of the bank in my bright blue dressing gown (for it was not yet breakfast time), and she showed no fear, although she passed no more than 20 yards from me; indeed she gave me the most casual glance, although I was possibly the first human being she had ever seen. From her plumage I knew that she could only be at most three months old, and it was clear that, in her manner towards human beings, she thought it wise to follow the lead of her elders and betters. Since they showed no undue alarm, why should she?

A week later Anabel would come up to feed with the other geese within a few yards of me, and if in the course of the winter I managed to create some impression upon her, it was perhaps more the bucket of corn which I usually carried than my personal charm that created it.

In February the geese, which had wintered in their thousands in the neighbourhood, started their northward migration, and by mid-March no pinkfeet remained near the lighthouse save the twenty-nine pinioned ones – and Anabel. She seemed to find the sea pool, the salting grass, the shelter of the bank and the lighthouse, the fresh water pond, and, above all, a daily cropful of corn, exactly to her liking. From time to time she would fly round, but she did not seem particularly restless, and I had gradually come to think that she would probably stay right through the summer. Then, on the morning of 16 May, when I went out to feed the birds, there was no sign of Anabel. During the night she had slipped away and set off northwards to catch up with the great flocks which must already have left Scotland for the far north.

Greenland, Spitzbergen, and Iceland, the breeding grounds of all the pinkfeet in the world, are dangerous places for a single goose. There are Arctic foxes, and falcons, and men, for all of whom a goose is just a very good meal. As October began I became apprehensive. There were also the dangers of the early autumn to be overcome, when the geese are stubbling in Scotland, and later in Yorkshire; a hundred possible fates might have overtaken Anabel. But none of them had, and, at noon on 9 October 1937, I heard her shout high up in a dappled autumn sky. She was a tiny speck when I first saw her, almost straight above me, and with bowed wings she hurtled downwards. She came in confidently, without circling at all, and settled at the foot of the bank 20 yards from where I stood with my bucket of corn. I called to her and she walked straight up to me. Any doubt which I might have had that it was indeed Anabel was at once dispelled. There she stood, a plump little round person, with her queer angular forehead, her unusually pink bill pattern, and the few white feathers at its base. To me she was as recognisable as a stray sheep to a shepherd, or a stray hound to a huntsman.

So Anabel was back; she had been away for four months and twenty-four days. The very day she came, my friend Michael was broadcasting a talk about geese and ducks on his marsh on the west coast. I sent him a telegram and at the end of his talk he read it out and told listeners about Anabel. This was less than seven hours after her arrival. A week later I chanced to be broadcasting, and was able to tell the story of her return more fully, so that Anabel became quite well known. She stayed again through the winter, with the birds at the lighthouse. Several other wild pinkfeet came in to join the throng, sometimes singly, sometimes in pairs and occasionally in small bunches. Anabel often flew round with them, but she knew that the lighthouse enclosure was her real home. When the others went off, as most of them did, with the departure of the main mass of the pinkfeet in February, Anabel stayed behind and came up to feed when she was called, with the pinioned birds. Although she led a sheltered life, safe from all enemies and with regular meals, hers was not a dull one. With her lived 150 wild geese from all parts of the world, to say nothing of a crowd of ducks. Many of her neighbours were not at all easy to get on with; disagreements, quarrels, and occasionally (though, of course, only amongst the more ill-bred members of the community) there were even fights, but there were compensations too, for few pinkfeet, after all, have had the opportunity of hobnobbing with an Emperor goose.

I was ready for Anabel's departure in the spring, and just as she had gone the year before, again she slipped away during the night, and on the morning of 10 May her familiar dumpy figure was nowhere to be seen.

During the last week or two Anabel's companion on her short flights round about the lighthouse had been a big gander who had been caught in a flight net a year and a half before. Instead of being permanently pinioned in the usual way, after which the flight feathers of one wing can never grow again, he had only had the feathers clipped. In the autumn he moulted out the cut feather bases and grew fresh feathers so that he could fly again. In spite of that, however, he stayed all winter through, and on the morning of 10 May, the morning after Anabel's departure, he came to feed as usual. He has spent the whole summer at the lighthouse. It seems that the interruption of his migration for one season was enough, in his case, to make the migratory urge less strong than the urge to remain with the flock. With Anabel it was the other way round, although the migratory urge did not become strong enough to make her leave until mid-May, more than two months after the big flocks had gone northwards. Geese as individuals are very variable in character and it may be that, although the big gander decided to spend the summer at the lighthouse, others of his kind in like circumstances would have eloped with Anabel.

It is almost certain that geese do not usually breed until they are three years old, and Anabel is only two. If she survives this summer's dangers in the far north, and next summer's too, then in October 1939, perhaps, she will bring her first family with her to spend the winter in the lighthouse pen.

From *Wild Chorus*

13

The Mysterious Sense of Direction

A pigeon is put into a basket and sent away by train. A hundred miles from its home loft it is released and three hours later it is home again. A swallow is caught at its nest and a tiny aluminium ring attached to its leg; in the autumn it flies southward to Africa and returns in the spring to the very same nest in the very same barn, and is recognised by the number on the ring. A Manx shearwater taken from its nesting burrow on the Welsh island of Skokholm and released at Venice is back in the burrow in ten days. A dog or a cat taken by train to the seaside disappears but is found again on the doorstep at home when the holiday is over.

How can these creatures find their way? How do they know in which direction to set off? How did Anabel return so surely to the lighthouse?

Anabel was a young wild pink-footed goose who spent a winter, in those far-off days between the wars, staying voluntarily with some tame pink-footed geese which lived in the enclosure round my lighthouse home on the Wash. It was in May that the migratory instinct finally overcame the instinct which directed her to stay with the little flock of tame pinioned ones, and she disappeared. Greenland, Spitzbergen and Iceland are the breeding grounds of the pinkfeet and Anabel's summer must have been spent in one or other of these northern countries.

I shall not easily forget the thrill on that October morning four months and twenty-four days later when I heard her voice high in the sky, and watched her circle and settle and come up to my feet to be fed. I marvelled then as I have marvelled many times since, at the strange power which brought Anabel back so unerringly to the home which she had found a year before.

In the history of scientific discovery certain mysteries and unexplained things have always made a special appeal to the imagination of mankind. In the field of natural history orientation is perhaps the most fascinating mystery still unsolved. So far, in spite of a recent and at first sight attractive theory, science has not been able to produce a satisfactory explanation of the phenomenon. Scientists do not yet know *how it is done.*

Each year millions of creatures – birds, mammals, fishes, insects – migrate hundreds, even thousands of miles, many of them finding their way with considerable, if not perfect, accuracy. And apart from migration many animals seem to possess a well-developed sense of direction for independent and individual movements – such as those of the homing pigeon; of the shearwater from Skokholm; of the hare which returned more than 600 miles to its home farm in Hungary, including a crossing of the Danube; of the salmon which find their way back into the rivers in which they were born. Even among human beings we speak of a good or bad sense of direction. This may only spring from good or bad powers of detailed observation, but alternatively it may prove to be a subconscious remnant of an additional sense still more or less highly developed among all these animals.

Is it safe to assume that the mechanism which finds the answer for them in their individual movements is, at least in principle, the same as the mechanism which takes the swallows on their migratory flight to Africa and back? I think that it is. I believe that the methods of orientation used by animals for homing will prove to be the same as those used for migration.

The most striking examples of this strange capacity are to be found in the bird world and it is therefore among ornithologists that the most extensive work on the problem has been done, and although the mystery is not solved its solution seems at the moment to be very near.

How far have the scientists got and how do the different theories stand? The schools of thought can be fairly sharply divided into two – those who believe that the powers of orientation will be shown ultimately to be derived from the known senses, and those who believe that some special sense or senses as yet unrecognised will provide the answer. Most of the recent theories have come from the second category and those who back the known senses (and principal among them the sense of sight) have for the most part been cast in the role of 'debunkers'.

Their argument goes on these lines. A bird may be born with an instinct to fly towards the midday sun as the days get shorter in the autumn. This is no more remarkable than many of the accepted instincts which are known to be inherited, as, for example, that a young grebe should have an instinct within a quarter of an hour of hatching to go to its mother's tail where it can then climb on to her back and be protected by the feathers of her wings, or that the newly hatched cuckoo should undertake the extremely difficult and arduous task of ejecting the eggs or young of its foster parents from the nest.

In the northern hemisphere the sun is in the southern half of the sky and therefore gives an indication of the direction in which to fly. The position of the moon, and even of certain groups of stars, might continue to give direction at night. An accurate sense of position might be obtained by a correct estimate of the height of the sun correlated with a sense of time, which birds are known to possess, and which is shown by the extraordinary regularity of their time of starting to sing in the morning. This time sense is recognised in man and may explain the capacity, which many people claim, to wake at a given time without the assistance of an alarm clock. Bees and ants are now known to use the position of the sun and its relation to various landmarks as a means of finding their way back to the hive or nest.

Into this theory of a general migration line on a particular bearing and orientated by heavenly bodies can be fitted the experiments of the German ornithologist Ruppell, who found that young hooded crows, which migrate on a line north-east and south-west along the southern shore of the Baltic, continued to use a parallel line if captured and displaced several hundred miles north or south.

The phenomena of migration alone could perhaps be explained by such a comparatively simple theory connected with the sense of sight; but what of homing? The argument here begins with an area of familiar territory around the bird's home from any part of which it could quickly return to the nest (or loft, in the case of pigeons). If a number of birds were released at some distance away and radiated evenly in all directions from the point of release a proportion would hit this area of familiar territory. The percentage would depend on the angle subtended by a familiar territory at the point of release. But if the birds, instead of radiating, were to perform an even spiral outwards from the point of release they would all in due course hit the familiar territory. It is not suggested that either of these things happens in this regular manner, but it is pointed out that a modified form is possible in view of the fact that birds are not infallible and that by no means 100 per cent of homing birds actually get home. In 1948 two Canadian scientists, Griffin and Hock, published an account of some experiments in which they released some gannets which had been taken from an island in the Gulf of St Lawrence to a point about 100 miles inland. The gannets were then followed, at a respectful distance, by the ornithologists in a helicopter. About 60 per cent of the birds eventually got home to their island, but the initial directions which they took were apparently at random and the tracks followed by the birds did not suggest any innate sense of direction.

Not only is the percentage of released birds which reach

home very significant, but also the length of time which they take to do it. The gannets averaged about 100 miles per day; and Griffin has shown that in this and a number of other homing experiments the percentage of recoveries and the average speeds on the journey are not inconsistent with what he calls 'spiral exploration'.

Many other suggestions have been put forward in order to support the theory that the sense of sight is the key to the power of orientation. Professor V. C. Wynne-Edwards of Aberdeen University has suggested that the accurate time sense of birds might detect the differences between the times of sunrise and sunset which change rapidly if you travel east or west. For every 100 miles in these latitudes the difference is about ten minutes. James Fisher has ingeniously suggested the comparison between a bird trying to find its way and a man in a maze. If, he says, you accept the principle of always turning in one direction it will probably bring you to the middle or to the beginning of the maze, but if it brings you back to a place at which you have previously been, you should take a new turn and then continue as before. You can do this in a maze because mazes have walls. For a bird, so Fisher's suggestion goes, the principles of geography – coasts, rivers, mountain ranges – take the place of the walls of the maze.

But although these explanations may cover a large number of the recorded phenomena of homing they do not, as their adherents would be the first to admit, explain everything. For instance, it is well known that birds immediately after release circle around gaining height, and that in a large proportion of cases they set off, after three or four circles, in the correct direction for home, irrespective of the topography of the neighbourhood and often without a sight of sun, moon, or stars. Furthermore many of them can find their way in a fog and at night as well as they can by day, although this is not the case with pigeons which are by no means the best homers among birds. On the other hand pigeons are moderate homers which are easy to keep and to tame and which breed freely in artificial conditions. It is for this reason that they are used for carrying messages and for racing. Since they are descended from the rock pigeon which is a non-migratory species, it is perhaps curious that their homing faculties are as good as they are.

If the stimuli of the known sense do not provide an adequate explanation of the homing performance of birds, what are the theories which postulate the possession of an extra sense or senses? Most of them are based upon a sensitivity to the earth's magnetic field and in this different observers have obtained different results on two very important pieces of evidence about which further experiments should be performed. First of all some scientists have stated that powerful wireless transmissions have upset the orientation of pigeons, while others have been unable to find any confirmation of this; and secondly, a number of experiments have been carried out in which birds have been released carrying magnets of sufficient strength to 'drown' the effects of the earth's magnetic field. In some of these experiments results indicate that the magnets may have had some adverse effects on the homing capacities of the birds, but it appears that in all such cases the results were not really conclusive, as insufficient 'control' experiments were carried out at the same time. Other observers have tried in vain to show any significant difference between the performance of a bird carrying a magnet and one carrying a small piece of non-magnetic metal of equal size and weight.

In any event an appreciation of the earth's magnetic field – a built-in compass, as it were – would not be enough to fix a bird's position on the earth's surface, for a compass is no good to a man if he has not a map, a chronometer and a sextant. He must know where he is before he can tell what course to steer to reach his objective.

It was at this point that Ising's theory of orientation by an appreciation of the Coriolis forces of relative momentum was put forward in 1945. Professor Ising's work was entirely theoretical and it was taken a stage further by Professor H. L. Yeagley of the Department of Physics of Pennsylvania State College who, in conjunction with the US Army Signal Corps, put the theory to the test. Yeagley's hypothesis was a combination of previous theories of magnetic reception with the new suggestion that birds might detect the Coriolis force due to the rotation of the earth. What is Coriolis force? If you throw a cricket ball out of the window of a moving train and at right angles to the line, it does not follow a path at right angles to the line, but rather a diagonal path due to the momentum imparted by the movement of the train. If instead of being in a train you imagine yourself standing at the Equator and facing northward you will be travelling at the speed of the rotation of the earth. Since the earth's circumference is 25,000 miles and it makes one revolution in 24 hours the speed is a little more than 1,000 miles per hour. North or south from the Equator your speed will be reduced until, as you reach the poles, it is nothing at all. If, as you stand at the Equator, you fire a rifle bullet, instead of throwing a cricket ball, in a due northerly direction it goes from a part of the earth's surface travelling at 1,000 mph to one which is only travelling at, say, 999.99 mph and it finishes up, as it were, a little farther ahead than the parallel of longitude along which it was fired; the rifle bullet drifts to the right. This is the effect of Coriolis force due to the earth's rotation. High speed aircraft find it necessary to make corrections on the courses steered on northerly or southerly bearings in order to compensate for this effect.

Ising's theory was that the semi-circular canals of the inner ear – the balancing mechanism of the bird – might be sufficiently sensitive to detect the Coriolis force.

Yeagley superimposed a magnetic sense upon the Coriolis factor and pointed out that a kind of grid could be built up based upon the two possible stimuli. Since the lines of latitude and longitude are related to the earth's rotational axis, the lines of latitude will also represent lines of equal intensity of the Coriolis force. On the other hand the lines of equal intensity of magnetic field will be centred upon the magnetic rather than the true North and South poles, and might be termed 'magnetic parallels' as opposed to parallels of latitude. If these magnetic parallels are superimposed upon the parallels of latitude the result is a grid-work of two systems of concentric rings which cross each other and which give an exact position on the earth's surface. If the bird could detect these two stimuli together it could fix its position precisely. But any particular magnetic parallel crosses a parallel of latitude twice, although the points may be many hundreds of miles apart. Such points are found on either side of the line of longitude on which the magnetic pole lies, and this passes more or less down the centre of the North American continent. On one side of this line the pattern of both magnetic and Coriolis intensity is, as it were, the mirror image of the intensities on the other side, and any point will have on the opposite side what Yeagley has called a 'conjugate point' at which the magnetic and Coriolis intensities are identical with those at the original point. When Yeagley decided to put his theory to the test with homing pigeons he found that the 'conjugate point' to the Pennsylvania State College, where his pigeons lived, was 1,100 miles away at the town of Kearney in Nebraska. If the pigeons were trained to return to a special loft at State College, Penn., and if the loft and pigeons were then transported to Nebraska and the pigeons released at normal distances from the loft (between 25 and 75 miles) they should, if the theory was correct, return to the loft, at Kearney instead of attempting the long journey across the continent to Pennsylvania.

Between 1942 and 1945 the theory was tested and the results, published in 1947, seemed to show that the birds were trying to return to the conjugate point instead of to their original home. For a while the mystery of orientation seemed to have been solved. But then, as scientists from all parts of the world began to study the details of the Kearney experiments, doubt crept back into their minds.

These doubts were crystallised at a most important meeting held at the Linnean Society of London on 13 May 1948. At this meeting the principal speakers were two scientists from Cambridge University – Dr W. H. Thorpe, the distinguished zoologist and Dr D. H. Wilkinson, a brilliant young physicist from the Cavendish Laboratory. Thorpe and Wilkinson attacked Yeagley's theory from two entirely different quarters and virtually demolished it.

Thorpe pointed out that the experiments were not conducted in a conclusive manner. The mobile lofts, each painted bright yellow and surmounted by a captive balloon 150 feet above it, were taken to Kearney and the birds were apparently allowed to remain in the lofts for one day in order to 'rest and acclimatise themselves to their surroundings' before being sent out in various directions for release. In spite of this 'rest day' and the captive balloons, only three pigeons out of five hundred actually returned to the lofts. The results, therefore, were chiefly based on a number of recoveries in the surrounding country which were judged to have indicated that the birds were *trying* to get back. These results were obtained by a method of plotting which Yeagley called 'the combined flight vector'. The fallacy of this form of measurement can be shown as follows: if six pigeons are released at 10 miles from the loft and one of them flies directly towards home, but proceeds past it and continues for ten times the distance, and if the other five then fly in the opposite direction for less than ten miles, the combined flight vector will indicate that all six flew in the right direction for approximately the right distance. Finally no bright yellow lofts with their captive balloons were taken to some quite different place which was not a conjugate point in order to ascertain whether the pigeons at Kearney, Nebraska, were doing something different from any pigeons released at random from any mobile loft anywhere.

Wilkinson attacked Yeagley's theory from the point of view of the physicist. Could the effects of magnetism or Coriolis possibly be large enough to be detected by any sensory mechanism in birds? He was almost certain that they could not and his arguments convinced a gathering of Britain's most distinguished ornithologists. He considered first the magnetic effects and ruled out the possibility that a bird could respond directly to the magnetic forces which come into play when non-magnetic matter is placed in a field. This would require an organ sensitive enough to detect a change of less than 0.005 gauss and pigeons have been subjected to magnetic fields of about 1,000 gauss without any visible reaction. It seems likely that any such sensitive mechanism would cause a visible reaction on the part of the bird when subjected to the shock of an application some 200,000 times greater than that which the mechanism was normally called upon to detect.

But there are two other ways in which a bird might be sensitive to a magnetic field, and Yeagley had put forward one of them as the basis of his theory. He had suggested that the bird might, in effect, be a conductor and that if moved in the earth's magnetic field a potential difference would be induced between the two ends of the conductor which could be detected by the bird. No current, however, is created, and Wilkinson showed that a bird accelerating from rest to 40 mph would have to make an electrostatic measurement of the order of one millionth of a volt. Since the most accurate man-made instrument for making such measurements – the cathode ray oscilloscope – cannot detect differences of less than about 1⁄10 volt, and since any such minute measurements would be hopelessly upset by the ordinary effects of atmospheric electricity, whose background intensity could not possibly be gauged by the bird, Wilkinson concluded that 'the induction effect as conceived by Yeagley is not operative'.

Yet another possible means of detecting the magnetic field might be available to the bird, however, if its anatomy contained a conducting loop which oscillated in the field. This would create an alternating current and the measurement of that current might be easier for the bird than a purely electrostatic measurement. But Wilkinson showed that this current must be measured to an accuracy of 10^{-10} amps and that in view of the much bigger currents of physiological origin which are present in living matter, it was, to say the least, extremely improbable that birds could make the desired measurement of the earth's magnetic field.

Finally Wilkinson turned to the computation of Coriolis forces. The effect of the Coriolis forces due to the earth's rotation could only be felt as a minute deflection of the downward pull of gravity. Its strength, in these latitudes, is less than $\frac{1}{6000}$ of the gravitational force itself and the angle of the deflection would be less than 1 minute. In addition the effect would be masked by the irregular influence upon the vertical component of the force of gravity of land masses such as mountain ranges and also of the centrifugal component. But this is not all, since the Coriolis forces due to the rotation of the earth could only be detected if the bird's course and speed were almost impossibly true. Any slight alteration, of course, would introduce Coriolis forces not related to the earth's rotation but to the change of course. Thus a bird would have to fly to an accuracy of $\frac{1}{50}$ of an inch in 100 feet.

Ising's conception of the mechanism involved in measuring Coriolis forces included a more complicated principle – that the forces caused a swirling in the fluid of the semi-circular canals of a bird of which the bird could become conscious. It has the additional significance that the method could be used when the bird is at rest; but once again the effect can be shown to be so minute as to make an extremely improbable method by which a bird could determine latitude.

Both the magnetic and Coriolis effects vary directly as the speed of flight. Thus a change from 40 mph to 39 mph would be equal to a geographical displacement of 150 miles. Thorpe and Wilkinson have shown conclusively that it is at least very improbable that the Yeagley theory is the answer to the problem, more particularly since the normal perceptions of animals – as, for example, in changes of intensity in light or sound – do not register such minute changes as would be necessary to make use of the Coriolis forces and the earth's magnetic field. But they have not shown that such sensitivity is impossible, and it may well develop ultimately that some part or parts of the theories of Ising and Yeagley will be found to hold good.

Meanwhile what other hypotheses are available? Wilkinson is thrown back to a visual explanation again. If random search or spiral exploration will not cover all the known facts, he suggests, might not these principles be materially assisted by an appreciation of latitude derived by a glimpse of the sun, correlated with the time sense which has been accepted? If the height of the sun could be estimated to within one diameter of the sun itself, then the necessary accuracy could be achieved.

In my view this is not enough. But there are other theories and we must consider in detail the most interesting experiments with swallows which have been carried out by two Polish scientists, Professors Wojtusiak and Wodzicki. Here are the important things which they discovered, the clues, as it were, from which they set forth once more to solve the mystery.

Swallows were taken from their meeting sites and released at various distances and in various directions from home. When released they circled once or twice and then set off. In two thirds of the cases the birds set off in the right direction. Some of the others began by following a railway line in the wrong direction. The birds returned as easily from any point of the compass. They returned almost as quickly through rain, thunderstorm and fog and their speed was reduced by only half at night, although swallows are not normally nocturnal. The farther the birds were taken away the higher the speed of return up to a distance of about 75 miles from which they returned at an average of 22 mph. At greater distances the speed was much reduced, but remained more or less constant at about 7 mph. Four swallows were tried over the same course a second time. One took a fraction longer on the second run, one did it in half the time, and the other two took about a quarter of the time. (This was particularly interesting in view of the training which is normally given to racing pigeons over shorter courses on the same bearing before a long distance flight.) Finally experiments were carried out with house sparrows and it was found that they could not home over greater distances than about 6 miles.

The two Polish scientists clung to the fact that the swallows could so often set off correctly both by day and by night, in fair weather or in fog. Here was evidently something outside the range of the conscious perceptions of the senses of man. Professor Wojtusiak put forward the suggestion that birds might have a visual perception of electro-magnetic radiation, for it must be something invisible to man, but existing in darkness and capable of passing through fog. Infra red rays have those properties. In support of this theory he called attention to some curious orange or red fat globules found in the retina of birds' eyes and also in those of terrapins and tortoises. If this should enable them to detect infra red radiations, then by the power of sight they could distinguish between warm masses and cold, between land and water, between the brightness of the south and the darkness of the north. The professor began his experiments at once, and soon showed that tortoises gave preference to areas illuminated with infra red rays, which may well explain the extraordinary manner in which water tortoises are able to find their way to the nearest water even if it is out of sight. The experiments extended to birds, and pigeons could apparently be shown to choose to feed from a dish lit by infra red in preference to one lit by an equal intensity of visible light, even though the positions of the dishes and the lights were frequently reversed.

But to set against these results are some researches by Hecht and Pirenne into the sight of owls. These two scientists failed to find any evidence that the eyes of the long eared owl were sensitive to infra red radiation. Although owls are for the most part non-migratory, yet the possession of a sense which would react to rays which are present at night would quite obviously be of the greatest advantage to a nocturnal bird. If such a mechanism were present in any birds, it is strange that it has not been developed in the highly specialised eyes of owls.

All these experiments go on and the evidence mounts up, but the answer is not found. To me it seems that the anatomists might now make a useful contribution. Birds which are known to be the most outstanding homers should be studied afresh and compared to those which are bad homers in order to see if any part of the brain, nervous system, or sense organs can be shown to have become more highly developed in those birds with the best sense of direction.

I am convinced that the birds cannot keep their secret much longer. Soon an irrefutable theory must emerge. Soon we shall know definitely how the swallows find their barns, how the great shearwaters scattered over the Atlantic Ocean find the tiny islands of Tristan da Cunha, their only breeding place, how Anabel found her way back to the lighthouse. Till then the mystery remains unsolved.

Much new research has taken place since that time which indicates that orientation is entirely based on the position of the heavenly bodies – the sun, moon and stars. In 1949 several solutions had been proposed, and were still to be proved or disproved. In nature every mystery solved unfolds new mysteries. We still have much to learn about the navigation and orientation of living organisms.

From the *London Mystery Magazine*, 1949

14

The Aura

There is a peculiar aura that surrounds in my mind anything and everything to do with wild geese. That I am not alone in this strange madness, I am sure; indeed, it is a catching complaint, and I hardly know any who have been able to resist its ravages, when once they have been exposed to infection. It is difficult to know just why this should be so. It is perhaps a matter both of quality and quantity.

I wish it were possible accurately to estimate numbers after they have reached the thousands.

I remember an afternoon at the end of September when a great gathering of geese were sitting on a big grass marsh. All day we had watched them struggling in in bunches of half a dozen to a dozen – tiny specks in the sky suddenly hurtling downward to settle on the marsh. They had done it all the day before, and the day before that too – arriving from Spitzbergen and Iceland and Greenland.

Some of them flew out and settled on the sand, and we tried to estimate their numbers. We counted and multiplied, counted and multiplied, starting first at one end and then at the other. Eight thousand was our estimate after half an hour of eye-straining through a field glass.

And then suddenly behind us a roar broke out, and the whole surface of the marsh seemed to rise into the air. A black cloud of geese, which conveyed just the same oppressiveness as an approaching rainstorm, moved out over the sand where sat the ones we had been counting. They did not settle with them, however, but stretched away down the crest of the high sand until those that pitched farthest were only visible as they turned to head the wind; fully two miles of solid pink-footed geese.

It was idle to return to our futile estimates. We could only gasp and murmur that our eight thousand were but a quarter of them.

Is it possible that twenty thousand geese made up that black line which stretched as far as the eye could reach along the high sand?

Perhaps there were half of the pink-footed geese that exist in the world here before us. Nearly all of them winter in England and Scotland, except for the few flocks which go to Holland. They breed in Spitzbergen and Greenland and a few in Iceland, and although it is possible to see vast numbers together, yet in their world distribution the pinkfeet are a tiny species of probably no more than fifty thousand individuals, possibly much less.

When they first come south on migration, they collect on this particular marsh, but after a fortnight they split up and go to other marshes and estuaries to spend the winter.

However many there may have been on that September afternoon, it was a sight and sound that must have thrilled the hardest heart. In this case perhaps it was their very numbers, or the volume of sound, or the mystery of their arrival from Arctic regions on the very date upon which they had arrived every year for who knows how long; perhaps it was the thought of so many great birds together, for pinkfeet are more than 5 feet from wing tip to wing tip; or the thought that, although the flocks are so big, yet the places they come to are so few – or perhaps it was the combination of all these things.

But probably the chief reason why wild geese hold such a peculiar fascination is because of their wariness. They are so difficult to be near, that being near them itself is thrilling, whether one is painting them or shooting them or photographing them, or catching them alive in a net. If you have been within 5 yards of unsuspecting wild geese you have achieved something, and that achievement alone has its special thrill. It is the knowledge that you have outwitted, not one, but perhaps a thousand *very wily creatures*. And there is another thing that makes wild-goose chasing so good. If you go after wild geese you will assuredly go into beautiful wild places at the most beautiful times of day – at dawn, or dusk, or moonrise – and, best of all, you will hear them call.

I have been many hundreds of times on a certain marsh in Norfolk, but I see it now as I have only seen it once or twice each season. And I hear the geese calling – a music of indescribable beauty and wildness, and fitness for the flat marsh which is their home. The moon has just risen, very large and orange over the sea, and the tide is high, half covering the salting, and filling the creeks right up to the sea-wall. There are a few wigeon calling as they fly along the shore, and away to the west a big pack of knots and dunlins twitter incessantly. Just an occasional call note reminds one that the geese are there waiting for enough light to come in and feed on the potato fields over the bank.

Suddenly there is a little burst of calling – the first ones are up – they're coming.

From *Morning Flight*

15

European Travels

In the middle of Hungary is the great grass plain called the Hortobágy. In summer it is a boundless expanse of green stretching as far as the eye can see in every direction from the little village of Nagyhortobágy; and in the winter it is a boundless expanse of white after the snowfalls in November. But in between, in spring and autumn, it is grey with a carpet of wild geese.

I went there first in March 1936, when the geese were migrating northwards towards their breeding grounds, and this is a letter which I wrote home to Michael, of my first impressions of the great gathering place.

The Csarda,
Nagyhortobágy, Debrecen

My dear Michael,

There really are several geese here. It seems of little consequence whether there are a hundred thousand or two hundred. So far I have only actually estimated sixty thousand at once.

We first saw the geese from the train. They were close enough to the railway to see that they were, as I had expected, mainly whitefronts; black bars plainly visible as the very nearest of them rose at about 100 yards. What we saw from the train I estimated at fifteen thousand – five rather scattered and the other ten all in one mass.

It was rather fun because we started watching like cats for the first sign of geese soon after we left Budapest three hours before! The first we saw were a long way away, about two hundred in the air. Then as we came to the green marsh we could see little dots of twos and threes flighting. Then all of a sudden we were amongst them, on both sides of the railway – at first scattered and then gradually thickening till the climax by a little copse where they looked like Abel Chapman's famous 'serried ranks'.

Well, the place is an enormous grass plain about thirty miles by twenty and bigger. At one side runs a river – the Hortobágy (which is pronounced Hortobarge, just like a canal barge). It is quite a small river – about as wide as the Backs at Cambridge. The marsh itself is so enormous as makes no matter. You can't see any edge to it at all. The surface is lovely green grass, very like Brogden, with pools all over it, only some of the pools are pretty big and more permanent, with little cut brows round them, but all quite shallow. There are no ditches or drains at all.

The landscape is broken here and there by a tiny copse (very few), and here and there by a sort of cowhouse and quite a lot by low green mounds which are relics of ancient civilisation.

Everywhere there are wells with great 30-foot lever arms for drawing water for the beast in summer.

The csarda (inn) itself is in the middle of the Puszta (? spelling but it's pronounced Pooshta, and anyway it only means 'plain'), about a quarter-mile from the railway which goes (at no great speed) through the middle of the said Pooshta east and west. So the csarda is right in the middle of the geese.

Well, we seem to have the most spare time sitting in our hides, which are square holes dug in the ground and full of straw, and so here I am in one, with a pen which will probably run dry at any moment. The morning flight is over and odd geese are flying about, and in between I write. But, nevertheless, I think I will go back to chronology.

It is very hard to write as I have nothing to write *on* and I have to keep looking up at every honk.

There are, of course, lots of other interesting birds – notably of prey. I saw several white-tailed eagles, a harrier, two rough-legged buzzards, and two red-footed falcons (I think, but they might have been merlins), ruffs and spotted shank, wigeon, teal, garganey, mallard, pintail, but ducks are not numerous.

There were about three lots of greylags during the flight, about twenty or thirty in each, but they are apparently commoner in autumn. There were also quite a lot of beans about. BUT, the *pièce de résistance*, the 'high spot' of the morning – three geese coming low on my right. One of them has a very short neck, it is very dark and small. As it passes there is a flash of chestnut. Eighty yards out and shining in the sun passes the first red-breasted goose I have ever seen alive. There are two lesser whitefronts with him. When he is past I can see that his tummy has much more white than the others – and so he goes off to be swallowed up in ten thousand geese which are sitting three-quarters of a mile away. I'm (blast! my pen's run out – now pencil). I'm sorry about the present tense, etc., but it *was* an event, wasn't it?

(Between us my live goose and I have made a wretched mess of this paper, because whenever I hear a goose coming the letter gets trodden into the mud!)

We started very early this morning, my second day. We came out to the north from the csarda, and since this wasn't a bit where the fifty thousand had been the day before – at least eight miles from it, in fact – I wasn't very hopeful, but one can't talk to the guides, because they only talk Hungarian, which is impossible, so there was nothing to do but wait and see.

I was put into a pit by a little pool and there were geese murmuring up wind in a semi-circle. It was a cloudless dawn and cold – thin ice on the splashes. There were a few odd birds about in the dawn. Suddenly there was a noise not only metaphorically but literally like a train. A grey mist appeared just above the horizon and all the geese in creation were in the air. They flew round and settled again. Then there was the same grey mist a mile away on the right and again on the left and then behind me – everywhere except just the track we had driven out along. There were certainly five thousand geese in each 'mist' and about five such mists within two miles of me all round, and this eight miles from where the masses all were yesterday. It's pretty safe to say you never see the same geese twice here! If there aren't more than a hundred thousand, there really and truly can't be less . . .

Almost immediately afterwards a big wave of geese came over. High above and behind them I heard a new noise, short but very squeaky, not unlike a wf. but fairly easily distinguishable. By now it was quite light and I grabbed the glasses. Were they? weren't they? – they were – thirteen of them. *Rufibranta ruficolis.* I didn't see much of the colour, but the shape was more compact and perfect than any other goose.

A little later I heard the noise again in the distance. It was another little bunch – sixteen redbreasts and one lesser wf. planing down with set wings from a great height. Their flight is almost reminiscent of a golden plover, so short are their necks.

Well, I can see the cart coming now, so I'll have to stop. I'll write again in a day or two.

Yours,
Peter

In the autumn of 1936 I went again to the great goose plain of the Hortobágy in Hungary. There was not such a great concentration of geese as I had seen there in the spring. Instead of a hundred thousand whitefronts there were only twenty thousand. There were fewer lesser whitefronts too, than in March, and I saw no red-breasted geese at all.

So we decided to go farther east and see if there were redbreasts to be found at the delta of the Danube on the Black Sea. Accordingly Tony and I took the train to Bucharest and then went by car down towards the delta. Our first stop was at Gropeni, a small village just above Braila. Here was a strip of grassy plain about a mile and a half wide which flanked the river, and which was sometimes flooded when the river rose in the spring. At this time of year it was dry except for a number of reedy ponds and a few larger expanses of water, with various channels and creeks leading out of them.

We arrived unfortunately on the very first day of the blizzard which heralded the continental winter.

Our first night was spent in great discomfort in a very primitive house on an island in the Danube, but next day we removed, complete with an army of guides and hangers-on, to the mainland again, and found another tiny house on the outskirts of the village of Gropeni. In this house I wrote of our adventures in a letter to John:

> ... We arrived in two cars at the same time as a perishing blizzard, and on the way we stalked about seventy whitefronts sitting near a ditch in the snow, on the edge of a grass marsh which borders the Danube on its west side ... Then over the Danube in a colossal Thames punt, with the most primitive sail. So far from being blue the Danube is the dirtiest brown that you can think of.
>
> In the evening we went to a deserted pool in a deserted wood where there were said to have been geese. In point of fact I believe a few pairs of greylags bred there in the

summer! Anyway we saw three ducks and no geese at all. This morning it still snowed and blew, and the army was unwilling to make the sea passage across the Danube in the heavy seas. Actually it was quite rough as the Danube is about three-quarters of a mile wide there and that is only a part of it.

The result was that we crossed in broad daylight, and since we had no idea in the blizzard where anything was (visibility two hundred yards), and since the army had considerably less, it was, as you can imagine, not much fun.

So we went home and removed to a new village on this side of the Danube, so that the army would not have to come plaintively to us as they did this morning and say, 'We have children, we have wives!' Anyhow, it was a shocking place, and I seldom remember to have spent a less comfortable night, having a wretched cold – no hot food at all, lots of mice – we haven't found the lice yet but expect them daily.

Here, however, it is more comfortable. This afternoon we went out in a sleigh, still perishing wind, but not actually snowing. We saw a tremendous lot of teal sitting (no doubt owing to the weather) as I've never yet seen teal sit, in bunches so tight that you couldn't believe they were birds, all along the edge of some enormous pools. We saw a peregrine take a pigeon from near the village. He looked grand lit up from below by the snow. Tony saw him actually grab: I only saw him carrying the parti-coloured pigeon.

We tried stalking the geese with one of our sleigh horses, because we had an asinine man with us who had spent the previous night telling us that it was the best method. Doubtless he had seen it successfully done, but, as I had expected, it was a case of once bit twice shy, for they got up at 150 yards. Ducks were very little tamer. Later the expedition developed into a round tour to see what we could see. First we saw about a hundred or more greylags sheltering in some reeds with some mallards (altogether two or three thousand I should think). Then we saw about six hundred whitefronts sitting so thick that one couldn't believe they were geese, mostly lying down and feeding, so as to keep out of the wind.

I looked at them with the glass, but it was hard to make them out, as they all looked black in the snow and there was a mist of powdered snow blowing past in front of them, but suddenly I saw some white streaks and there they were, seven little redbreasts. Part of the flock moved and when they settled again they were farther away and there were twelve redbreasts, and still two stayed with the near lot. We walked fairly close, but it was too cold to hold the glasses up for long. I tried with the horse again, but they got up at at least a hundred yards. As we left I took a last look at them over on the far grass where they had settled (a place where the wind had blown the snow mostly away). They all turned and walked down wind picking their way amongst the whitefronts – fourteen little redbreasts all in a row.

After that we saw about fifteen hundred more whitefronts, but not to peruse in detail, and quite a lot of distant geese (a woman has just brought in the dirtiest jug of water I have *ever* seen – not only is the water dull mud colour but the jug itself is covered with large splashes of congealed gravy and the rest of it is of such a dirtiness as staggers all belief. I may say it is intended for drinking water – washing water is entirely unheard of here).

—

It had snowed solidly since last night (it's now tonight – somewhere about Thursday). We went forth this morning but we might as well have saved ourselves the bother. It was getting light before the army was finally mobilised and broad daylight before we could get into position and then not where we meant to be by about two miles. It was snowing a bit and the ice would bear moderately, but eventually Tony fell through and got one boot full. The greylags were there and about two hundred whitefronts and hundreds of ducks, but they all went off when we arrived – very likely they had been there all night. Then the most terrific blizzard set in, with wind speed about 35 mph and visibility under one hundred yards – and so there was nothing to do but find the sleigh and come home – against the blizzard too. No bird would fly for choice in such weather, so there was no point in staying out.

Since we have been back it has blowed and snowed continuously. There is a stream of water a few hundred yards in front of the village and in lulls in the storm we can see teal in the water and about a hundred and twenty whitefronts feeding along the edge.

Well, it seems the continental winter is upon us. The blizzard has swept down from the steppes of Russia and here we are, as you might say, 'cetched'. The snow has drifted and all communications are severed and even the telegraph wires are down, so we may be here for a while yet, but I fear the geese won't be, as the snow's getting pretty deep. I'm having an ice sled made just in case it stops snowing, but I don't think it ever will.

—

Well, it did, and this morning it was clear but still blowing fit to perish. There were about two hundred and fifty geese feeding along a stream about half a mile away from our window and after breakfast five geese feeding along another stream about three hundred yards away. These five were beans and Tony set forth with a white horse to stalk them. Since there were numbers of peasants tending their boats and drawing water, etc., within about one hundred and fifty paces of them, there seemed a fair chance of success, but we had not reckoned with Pegasus, who behaved in a most unruly way. However, after waltzing round two or three times, rearing on his hind legs, and neighing loudly, he cantered off leaving Tony kneeling within about seventy yards of the astonished geese.

Soon after the affair of the beans, we set off in a sleigh. The snow was much deeper than yesterday. There were still lots of ducks but hardly any geese at all – about twenty 'lags where they had been before. There was too much snow on the ice to try the sledge, particularly as it had thawed and was slushy on top. Farther on we came upon ten geese feeding exactly where the redbreasts had been two days ago. They were whitefronts. Then we went miles and miles in the sleigh, drawn by two of the most miserable horses yet seen, and saw no geese at all. The snow had driven them all south. We went to another village and had some sardines and beer, and then eleven kilometres home and the road so full of snowdrifts that we had to go over the fields and we had to walk through snow a foot deep mostly, because the horses couldn't pull the sleigh *and* us. It took us about two and a half hours and was dark about halfway. We'd been about sixteen miles and seen about one hundred geese, and one other nice thing, just before we got to the far village – six great birds, slow flapping and high in the air, lit from below by the snow and looking very pale – bustards. Added to these we must have seen about fifteen white-tailed eagles, a merlin, a rough-legged buzzard, and a kite.

The snow has driven the geese away so much that tomorrow we are off – Tony home, as he has to be in London by Tuesday, and I farther down the delta to see what is to be seen down there.

—

Started in a steamer between Tulcea and Valcov
on the Danube – not so truly blue

My Dear John,

Things haven't got much brighter lately. The snow has pretty well spoiled everything. I packed off Tony to London (according to previous arrangement) about two days ago from Braila, which is a truly miserable spot, especially in a snowstorm. I went to the pictures and saw Mr Warner Oland in a Charlie Chan film. Then at 7 a.m. next morning, in a wretched snowstorm, I made the discovery that my Leica had been stolen. I have an Insurance taken out in Debrecen, so they may?? pay me for it. Anyway it wasn't too bright a start to a ship journey, which I had postponed till the morning so as to see the country. Visibility was about three hundred yards most of the time. I saw six hundred geese in one place. Arrived at Tulcea, I went to call on Dr Rettig, whom I knew by name as the local ornithologist. He was charming, and a good taxidermist. He had stuffed two redbreasts. Of course, he didn't believe I had seen fourteen at Gropeni.

At about midday we embarked on this little ship. The point is that there are known to be geese at or near Babadag, and believed to be geese at or near Valcov, which is pronounced Vallcough – to rhyme with Hall-cough – and which is also where caviare is caught!

Well, the snow is such that the thirty-five kilometres from Tulcea to Babadag could only be traversed in a sleigh or wagon, taking eight hours, whereas in a day or two it may be passable to autos if the snow goes on melting as fast as it is now. So the plan is to go by boat (which doesn't mind if it snows ink – much) and then, if that is no good, come back, by which time the road may be clear. So that is what we are now doing, Herr Lautner (my trusty guide) and I. On the way we have seen three lots of geese, about one hundred in each, and a good many ducks roosting afloat in the Danube. One lot only got up when the boat was about fifteen yards from them. I have seen some kites and a good many eagles and buzzards (rough-legged) and peregrines, and a very pale grey harrier which I think must be a pallid harrier. Teal and wigeon fairly common – one pintail drake. Anyway, there should be plenty of pelicans at Valcov, and I shall be quite amused to see them.

The snow has been wretched bad luck, as up till then there really must have been a lot of geese. All are agreed that they blackened and deafened respectively the sky and the ear!

Herr Lautner has just come to say that the Crystal Palace is burnt down; poor old Crystal Palace!

The engines have stopped so I suppose that means we're there – if only ten thousand geese are too, all will be well.

7:30 p.m. Three-quarters of an hour later. It seems as though they may be. We have arrived, installed our baggage, and I am now sitting in a nice little wine house (they don't have any beer here) surrounded by staring small boys, and eating chicken and rice. Near at hand is Herr Lautner cross-questioning all the fishermen, and from the frequence of the word 'göshti', I believe we've found some geese at last.

The people here mostly talk Russian. Quite a lot of them have snow on their boots! They sing, which is quite merry of them, but I do wish they didn't eat garlic. Also, they all play backgammon.

Here are some miscellaneous facts about Roumania in general:

There are lots of wolves. They ate six sheep at Gropeni the night before we left.

Hoodie crows are the commonest bird and take the place of sparrows in the villages. Rooks and jackdaws also live in the villages and towns, feeding in the streets. The only kind of sparrows I have seen are tree sparrows. I've also seen goldfinches and any amount of crested larks.

We all but got arrested at Gropeni because we went out without Herr Lautner and were unable to explain that our innumerable papers were at home.

A census of verminous insects in Roumania would look like elaborate astronomical data, whilst the insects themselves would, I am sure, if placed end to end, easily reach from here to the moon!

Here one can eat caviare with a tablespoon (only I haven't done so yet because I haven't been here long enough).

I am now in Bessarabia – so what do you think of that?

—

Twenty-four hours later and I sit again in the same little wine house. This town is a sort of Venice, full of canals, and it is also full of frontier outposts – soldiers, etc. Anyway we left at an unpunctual five from a canal outside our very door in a sort of coble – a large black boat with peaked ends rowed by two lusty young fishermen. It was a big boat, about thirty feet long, so as to brave the perils of the ocean.

After rowing I should say about three hours – eighteen kilometres, down a widish arm of the river flanked with reeds, we came to the Black Sea – only it was an ugly yellowish grey as it was blowing quite hard. There were hundreds of small boats fishing for sturgeons (actually about thirty, but they were all in a bunch and they looked a lot).

We turned north and went along the shore. The sea was quite calm as the wind was off the shore. The shore itself was reeds, colossal reeds. These delta reeds are something special. They are almost all fifteen feet high. I measured one of eighteen feet; and they are said to grow to over twenty feet in places. The water was very shallow near the edge (there are no tides) and presently we came round a corner and saw a very marvellous sight. There was an enormous bay, perhaps two miles across, and parts of it were shallow, with grey sandbars showing. There were masses of ducks, and there in the middle, and quite near, was a great flock of whoopers. There were seventy-five of them, a few young ones, but mostly white. They made a tremendous noise when we put them up. By now we were sailing and going fairly fast, so we got quite close. They got up and settled at the back of the bay with some more. I counted one hundred and seventeen in that lot with about sixty greylags amongst them. The ducks were mainly mallards and teal, but after that gadwall were the commonest. I saw dozens and dozens of them. There were shovelers and wigeon and three pintail drakes (presumably with their ducks) and shelducks, and about a dozen pochards and later four tufted ducks, and a pair of goosanders, and overhead was a white-tailed eagle, and at the back, over the reeds, were two species of harrier and rough-legged buzzards. There were about ten thousand ducks, and then further on on another spit were eighty more swans and about one thousand geese – all greylags to a bird. Some of the ducks were in shallow water and amongst them were two gulls – but even gulls in these parts might be interesting, I thought, so I looked at them with the glasses. They were avocets. Beyond were curlews. We passed the geese and came to a little fishing hut. There were four men preparing sturgeon lines. They catch them with a sort of spiller – a line with bare hooks (very large) hanging on short strings every foot or two along it. The hooks aren't baited at all, the principle being that the fish just runs into it by accident. It is attached to floats so that when one hook catches hold, the others coil round and also catch in the sturgeon and he's soon 'taffled up small'.

They only catch four or five in a year, each fisher, but they fetch about £30 each or more.

We sailed on northward and saw two drake goosanders shining in the sun, and about three hundred more geese, all 'lags again. They were all sleeping on the sand and clearly must have been feeding at night under the moon. But where do they feed? Local legend has it that they don't fly at all, but subsist on a kind of aquatic nut which grows in these parts, and is rather larger than a walnut, only with sharp spikes all over it. That I believe to be an old wives' tale! That leaves two possible explanations – one, that grass did grow on the sandbanks until the recent storms in which eight ships were sunk in the Black Sea (which is what the people say), and two, that somewhere inland from the reeds there is a feeding place for the geese. This last theory was backed up on the way home, when we saw three or four lots of 'lags going inland at colossal height in a north-westerly direction.

I think it is likely that I shall go away again, as the main point is the redbreasts and mere aren't any with the 'lags – and time is getting short and I like seeing as much as possible now that I am here. I shall probably go down to Babadag if the snow permits. And now, if I don't stop writing, I shall fall asleep where I sit, which will be a treat for my fellow diners. So good-night.

—

Hortobágyicsarda, Hungary

My Dear John,

Here I am back again and sitting in the csarda bar (because the season is really finished and the dining-room is closed down) and the time is about 5 p.m., and there is quite a lot to tell. The story is chiefly of entomological interest. Having travelled a good bit in Roumania I really must have improved the breeding stock of the vermin herds tremendously by the introduction of fresh blood into each neighbourhood.

After posting the letter to you on the ship and eating a teal which we had cooked for lunch, we went ashore at Tulcea at about midday and found that the snow was still blocking the road to Babadag, which you may remember is in the southern part of the Danube delta, to the exclusion of autos, and so the only way was five hours in a cart; and what a cart! I consider that five hours to be about as uncomfortable as any five hours of transit I have ever spent. As Herr Lautner remarked ten minutes after we started – it was enough to rattle the eyes out of one's head – and it went on for more than thirty kilometres. We passed over some fairly high hills with six foot snowdrifts still uncleared on the road. We had to go round them over the fields, which was the only comfortable part of the journey. There were the remains of war trenches still all along the road. Well, at the end we came to a marshy valley at the head of Lacul Babadag, which is quite a big lake. It was mostly frozen up at the top end, with big islands of high reeds and several hundred ducks sitting on the ice. In one place were about forty greylags sitting in the water. It was getting dark by the time we had found a house in the village of Zebil in which to live. And what a house when we did find it! One old man lived with his son and would be glad to put us up if we had some tobacco for him. He had only one room – it was in fact the house. The room had a big square stove of mud in the corner (as all houses in these parts do), and from it one half of the room was raised about eighteen inches like a sort of stage. Admirable for charades, but less attractive as a bed (with one big quilted blanket) for myself, Herr Lautner, the old man *and* his son. After supping off a little black bread and some pickled water melon it was time for bed. With the utmost difficulty I managed to prepare a jersey to cover the pillow without giving offence and eventually we all lay down in a row. Having always learnt that a flea was remarkable in that it could jump one hundred times its own height, comparable to a man jumping as high as the Eiffel Tower, I spent some time idly speculating on the precise height of a flea. But before very long it became painfully apparent that there was not room on the shelf for the four of us to be out of hopping range – so that was that.

Nobody could call me fussy – and I hadn't had any clothes off at all for over a week (the principle being that they had more trouble getting through if one kept well battened down), but I must say I found that night uncomfortable. Next morning Herr Lautner and I went out early, but we saw very little except for a fox running across the ice and some ducks in the open water in the middle and the forty geese which we walked into feeding on the shore of the lake. We heard quite a big lot of greylags over the other side of the lake.

We went back to a splendid breakfast of what remained of last night's black bread, still on the table and a good bit staler than it had been: and since it had begun to snow again I decided to quit. There's no doubt the geese *had* been there. One or two fields I walked over which were clear of snow were absolutely smothered in droppings. But I was fed up with being always too late.

We ordered a waggon to take us to Babadag. One the way we drove over the causeway. There were waggons and people going along this causeway all the time, but the ducks were crossing it – flying into the snowstorm only about twenty feet up – and streaming across. And so were a good many 'lags – a bit higher, but not much. Further on I suddenly saw a single whitefront sitting in a field. Then we saw four hundred 'lags sitting in a bay and, as we watched, a lot came in and settled on the grass quite near us, and a flight began. The flight passed over a reed bed and I was sure I could get under it unseen, and even if the flight was finished by the time I got there I would be in direct line if they were put up, and so close that they would be sure to pass over me. I was all but there, and the bunches were still swinging in over the reeds one hundred and fifty yards in front of me, when a blessed cart must needs go trundling past and out went all the geese. I waited for a while but they didn't come back, in spite of an eagle which tried to help. It drove some geese and ducks towards me, but the geese turned off. And so we went on and caught the train at Babadag and were in Bucuresti (Bucharest) at eleven and I went, just as I was – rubber boots, four or five days' beard, and a fortnight's fleas – to the best hotel, the Splendid Park.

Next morning I took a census of insect bites, but after counting three hundred I gave it up. Later in the day (after hearing news of the constitutional crisis) I took train to Puspokladany (if you can do that one first go off you're cleverer than I thought you were!) and thence to Debrecen. I got to Debrecen at five o'clock this morning and slept at the Queen of England Hotel until eight-thirty, and then old Nemeth drove me out to the csarda and so here I am. There is a little snow on the puzta and all the water is well frozen. The report was no geese and very few ducks. This afternoon I walked round the new reservoir.

There was a little patch of open water on the reservoir, and about one hundred and fifty ducks where there had been twenty thousands. But, although it's disappointing to see the place bare of birds, there's something rather awe-inspiring about the continental winter and the way it has come down with a north wind driving all before it.

However, having heard about the absence of geese, I was surprised to hear a roar over Borsus, and presently the first waves of birds arrived to roost on the ice of the reservoir. Altogether at least two thousand geese came

> to roost from Borsus, but they say that there are more at Holaskus, so I shall go there in the morning although it's much farther.
>
> I shall have to wait here until Wednesday. Nemeth Ur has sent away to the next village, ten miles away, for my favourite gypsy violinist, and he has just arrived by the train and is now playing 'Nem tudom ehn mivon vellem'. This really is rather a sweet place. Everyone treats one as if one had come home. Tremendous welcomes. When we left for Roumania a fortnight ago the gypsies brought their instruments out and played us into the car and away! And now they are playing my favourite tunes in honour of my return, including one they have composed about 'Scott Peter' and his hunt for 'vörös nyaku liba' (red-necked geese).
>
> The live geese are rather exciting. Since I have been away they have collected quite a lot. So far the nets have not caught anything, but the moon has a lot to do with that. But the pen contains sixteen geese – eleven whitefronts, one bean, and four kish lilliks – and on Thursday I shall set off for England with them all.

And on Thursday I did set off with the live geese in three large crates. At Budapest they were loaded into an aeroplane and we flew to Vienna. At Vienna there were reports of bad weather and fog ahead. I and the geese were the only passengers and the pilot told me that if the fog were bad we would not stop at Prague but go straight on to Leipzig.

But the fog was bad at Leipzig too, and we had not enough petrol to go on. We came down gingerly, dropping into the grey blanket of fog, saw a chimney-top flash by, and zoomed up again into the sunshine. Three times we tried unsuccessfully to get down, and then flew a few miles to a military airport at Erfurt where we found, as the wireless had reported, a thin patch in the mist and we landed safely. We refuelled there with four hundred gallons poured from tins, and then we took off again and flew to Cologne.

At Cologne the service was suspended. England was enshrouded in fog. There had been a bad accident at Croydon, with the loss of fourteen lives that morning, and the geese and I could go no further by air. There was a chance, they said, that I might catch the boat train to the Hook and be in London the next morning; and I was anxious to do this because the crates were small and the geese unfed, for I had been counting on a short journey. But I had so arranged my affairs that I had only just enough money for my taxi in London, so that I might obviate the currency restrictions on the journey across Europe. By the time the airline had refunded the exact railway fare necessary there was hardly time to catch the train. The bus hurtled from the airport to the station to the great discomfort of the other travellers, and accompanied by two airline porters I dashed in to buy my ticket. One of them, carrying my baggage, came with me, whilst the other took charge of the geese.

'You have luggage to register?' asked the booking clerk. 'Then you will have to pay extra for it.' 'Yes,' said I, 'but my train is due to leave, I will have to pay on the train.'

'You must pay now.'

'I have no money,' I replied, and threw down a few pfennigs which was all that I had. The porter pulled me by the arm. 'Never mind him,' he said, and we dashed up the steps on to the platform.

There was a train moving past the platform at a good speed. The other porter ran up. 'Your geese are in there – in the front van,' he shouted, and I could see that there was no chance of loading my baggage on to that fast-moving train. But the geese were going to the Hook: baggage or no baggage I must go with them. So I leapt on to the running board, shouted back to the porter, who said, 'All right, I'll send them after you', and pulled myself up into the train.

Once inside I took stock of my position. All that I had with me was a Hungarian fur-collared overcoat, a volume of short stories by Somerset Maugham, and a cheque book. I had no money and I had had no food since a roll at Vienna at ten in the morning. As I walked along the train and through the dining-car I was tortured by the smell of the dinner which my fellow travellers were already enjoying.

I appealed to the head waiter of the dining-car, but he had no suggestions to offer.

When you are hungry it is bad to think about food. I went and sat in a carriage farther up the train and thought about food for half an hour. I had just started to think of something else when the little bell sounded in the corridor and the attendant announced the second service. A few moments later an official came to inspect my passport.

'Are there any English people on the train?' I asked him.

'Yes, there are one or two.'

'Which of them would be most likely to lend me a pound?' and I tried hard to look as though I were not a 'confidence man'. But he looked so suspicious that I had to tell him my story and when he admitted that he had heard the geese honk as he passed through the luggage van I knew that I could count on his support.

He led me to a benevolent-looking Englishman who, when I had told my story over again, willingly cashed a cheque for me, and ten minutes later I was eating a hearty dinner.

From then onwards the journey went smoothly enough and by the following evening we had all arrived safely at the light-house – eleven whitefronts, four lesser whitefronts, one bean goose and I.

From *Wild Chorus*

—

Far out in the Transcaspian Steppe in northern Persia we found Atagel, in a hollow behind a low ridge. As we topped the crest, it lay below us, a shallow lake, brilliant blue in the pale sunlight of a winter's afternoon.

The surface of the water was dotted with birds, white dots and black dots and, near by, piebald dots which were shelducks up-ending close to the flat sandy shore. A little row of large and very white dots was a party of pelicans fishing out in the middle. The pink of their plumage made them look whiter than white, but at that distance not pink at all.

There was a family of whooper swans, two white and three grey; and then there were the ducks. They were mostly at the far side, where a little sluggish stream spread out into a marshy delta at the edge of the lake. Wigeon and teal were there in great packs, and there were gadwall and shovelers and red-crested pochards and smews.

Over the flat marsh flew a little party of greylag geese; but standing in the shallows at the edge of the open water was a much more exciting little dark line of birds. They were larger than the surrounding ducks and smaller than the greylags, and most of them were asleep.

I had come three thousand miles to find red-breasted geese, and here, less than a mile away, were fifty sleeping dots that were about the right size, yet too far to show any colour.

Then, as I sat gazing through the telescope, I heard, very dimly at first, the call of geese flying high somewhere away to the northward; it was the high, thin call of lesser white-fronted geese. Presently they came into sight, two hundred strong, flying in a magnificent V. When they saw Atagel they set their wings and started to glide. A fresh burst of calling came across the water and in that instant I knew that I had still to find the haunt of the elusive redbreasts, for they have quite a different, high-pitched barking call. The answer from the water was lesser whitefront language.

The formation broke, and some of the geese circled back and round, losing height. But most of them flew on, re-forming again, past us, and away south-westwards to the Caspian Sea.

Their chorus had died away in the distance by the time that the two small parties, which had broken away and circled downwards, finally joined the flock which stood in the shallows.

I sat for a long time wondering at the vastness of what I had seen. Behind me were the mountains of Persia, a snow-capped frieze stretching across the southern horizon. Somewhere between me and the mountains, in the corner of the Caspian Sea, was their journey's end, the southern limit of migration of the lesser whitefronts. In front lay the steppe whence they had come, unending dry grassy plains stretching away and away to Bokhara and Samarkand.

From their breeding grounds north of the tree line, two thousand miles away, they had come southwards, perhaps following the side of the Ural range, or perhaps the great River Obi, the Irtish, the Tobol, and then striking across the Kirgiz Plain to the Aral Sea.

From there they had come yet farther south, stopping here and there at some lake or river marsh on the way. Atagel was one of those stopping places, and I had seen some of them on the last stage of the great journey.

I followed that south-going skein towards the mountains, and, after hunting for a week, I found at last the lagoon where all the lesser white-fronted geese of the Caspian Sea collect in mid-winter.

From the foothills and the woods I rode out on to the marsh with a high heart. One of the great goose marshes of the world was before me, and far out at the edge of the lagoon smoky clouds of geese rose, circled shimmering in the mirage, and settled again in untold thousands; and a distant murmur like the high singing of gnats came in from the shore.

The horse's hoofs splashed in the pools and puddles, and the grass was gloriously green. Snipe jumped up in front of us, and an occasional pair of mallards clattered out of the rushes. Beyond a belt of dark green rushes were the nearest geese. I went forward on foot and then stalked to the edge of the rushes.

There were about three thousand on the short grass before me, and many more stretching away into the rushes beyond, and every one that I could see was a lesser white-fronted. Surely these exquisite and delicate little birds, with their golden ringed eyes, their tiny pink bills and smart white foreheads, their brilliant-orange legs and black-barred tummies, deserve a more romantic name. The Persian 'goz siah kuchik' means 'little black goose', to distinguish it from 'goz sefid' – 'white goose' – applied to the greylag. These names serve well enough in a land where the snow goose and the brent are unknown.

The natives of the nearby village of Kara Tappeh knew the red-breasted goose, but only as an occasional visitor in small bunches, among the hordes of lesser whitefronts. They called it 'Shahpasand goz', meaning literally 'worthy of the Shah', or, perhaps, 'Royal goose'.

'Kara Tappeh' means 'a black hillock', and the village stands proudly on the little mound in the middle of the marsh, with a characteristic wide-eaved house silhouetted on its very summit.

I stayed there in the house of the head man for a week, wandering on the great marsh. It is about two miles wide between the lagoon and the woods, which are full of enormous boars; tigers roam in these woods, and there are wild pheasants and porcupines. The marsh is about ten miles long, and on it there cannot have been fewer than thirty thousand lesser whitefronts, and I think there were twice as many.

Morning flight was always late, usually just after sunrise, and for half an hour the air was a network of skeins coming in high from the lagoon. As far as I could see in both directions were more and yet more, into the far distance. The black lines on the mud flats became gradually thinner until finally no geese were left at all. Out beyond where they had been the sun lit up the vast crowds of flamingos, and the mirage made them into a rectangular coral pink bar of amazing brilliance.

Behind, over the marsh the geese were flying about in little parties. Occasionally whole masses would rise, disturbed by a shepherd or an eagle or a gyrfalcon, and then settle again like a grey carpet.

I saw no red-breasted geese in the week that I was there, but I learnt the innermost thoughts of the lesser whitefronts on the green marsh at the foot of the snow mountains. It is a wild setting for a wild wild-goose.

—

Siah Derveshan is a typical Caspian village of thatched-roofed cottages built upon the bank of a river in the middle of the great rice swamps of North Persia.

I came to it on a February afternoon, and I liked it so much that I stayed for a fortnight

We had come by boat from the port of Pahlevi, and we were exploring the great lagoon, searching for red-breasted geese. We had been round the western end of the lagoon, rowing in and out of reed-fringed bays amongst pelicans, flamingos, cormorants, swans, and a host of smaller waterfowl, but the only geese that were to be seen were greylags and whitefronts. We travelled in two small bright blue row boats, in one of which were supplies and baggage, and in the other myself and my faithful Ismail, a Persian youth who spoke a few words of German, and who was, therefore, my interpreter. To propel these vessels came three sturdy boatmen.

It was on the third day that we came to the bay of the 'pig-puddles'. A kind of reedy grass had once grown in the bay, and this had been completely ploughed up by the snouts of wild pigs. There were perhaps two or three hundred acres of this grass, and not one square yard but had been turned over and laid flat by the pigs. The whole bay consisted of little puddles bounded by rolls of matted reed. In the puddles floated little brown seeds and to eat these seeds came thousands of ducks. It was a daylight feeding place, for at night it was almost deserted; but when we first came to the place at about noon there were perhaps fifteen thousand waterfowl in the bay. Mallards and pintails were there in equal numbers, with a sprinkling of other surface-feeding ducks, and amongst them a crowd of geese too. There were greylags and white-fronted, but none of the red-breasted geese for which I sought. Out on the open water of the lagoon were seventy Bewick's swans, and farther out still a party of whoopers. Away to the left sat a shimmering pink line of pelicans on a point of mud. The sun shone out of a clear sky and the trees across the bay were magnified by the mirage.

Behind the bay of pig-puddles was a forest of alders and willows and reeds. Parts of it were swampy and I was told that in these swamps wild ducks were caught at night by strange means; so I decided that this would be a good place to stay for a few days to investigate the stories I had heard.

Half a mile to the eastward was a little river, and a fisherman, whose boat was beached there, told us that there was a village hidden in the trees and that its name was Siah Derveshan.

I walked up a footpath which followed the bank of the river, here and there cutting off a corner and striking across meadowy park land. Ismail came with me and the boats came after us following the lazy curves of the river.

About a mile from the lagoon the land had imperceptibly risen, so that the river on our left was running between steep banks ten feet below us. On our right the woods were flooded and we could hear mallards and teal calling and occasionally see some of them in the few open patches of water that we passed.

Round the corner of the woods we came upon the thatched roofs of the village clustered around the fish trap which spanned the river.

In the village was a small bazaar round a closed courtyard. We entered by one gate and I suggested to Ismail that we should find the head man of the village and ask him if he knew of a house in which we could stay. But Ismail, who was only twenty and shy of disposition, was disinclined to embark upon this self-invitation. I suggested that we might ask for the owner of the nearest *Murdab*. *Murd* means dead and *ab* means water, and this was the name used for the swamps in which I had heard that the ducks were caught by using a flare at night. Under pressure, Ismail made inquiries, and ten minutes later, as I had expected, we were being led off up the river bank to the house of the chief duck-catcher, who told us that we might stay with him as long as we liked.

It was a lovely two-storey house with a wide-eaved thatch covering a wooden balcony, and our room was the one in which our host dried his tobacco crop. It had more windows than most Persian rooms, many of which, indeed, have none, and round the walls were neatly stacked bundles of tobacco leaves, for, as duck-catching was his winter livelihood, so tobacco-growing provided his summer income. There was no furniture in the room except a fine carpet and a little brazier of charcoal which was brought in soon after our arrival. Sitting, eating, and sleeping are all done at floor level in Persia.

At dusk I was led down through the village again and along the river bank. After half a mile we branched off a few yards into the wood and came, all at once, upon a boat lying in a narrow ditch. It was a tiny flat-bottomed craft with a peaked bow and stern, no more than 12 feet long and 2½ feet wide, and when we embarked there seemed to be horribly little freeboard. Propelling the punt by pole, my guide pushed along the ditch for fifty yards and then turned into another ditch which led at once to a small thatched mud hut. Here there was another punt with a strange curved hood over the bows like the hood of a cobra. In the hut were three more men, one of them our host, the chief duck-catcher or *murdabchi*.

We waited for darkness sitting around a wood fire in the middle of the hut. There was no chimney and the place was full of smoke.

Some time before I had heard about the method which the Caspian duck-catchers used; how they went forth at night with a flare and a gong and could pick the ducks up by hand or catch them with a butterfly net. I had not been able to believe all that I had heard, but how much of it was true I felt I should very shortly know, and as I sat waiting my excitement grew. So it was with a thrill of anticipation that I emerged from the four-foot-high door of the hut an hour later, my eyes streaming from the effects of the wood smoke. One of the men who followed me carried a small brass gong, hanging from an inverted L-shaped stick, in one hand, and a gong stick, padded with flax, in the other.

Outside the hut, by the light of a flaming torch, I could see two large hand-nets leaning against the roof. They were about 12 feet long and only about 2½ feet wide. Seven feet of their length was net, bounded by curved willow poles, and the other five feet was handle. At the tip the willow poles were drawn together by a string till their ends were only a foot apart. The net was of very small mesh, so that a sixpenny bit would hardly have passed through it.

One of these nets was selected by the *murdabchi* and I handled it for a moment. So beautifully was it balanced that it seemed to have practically no weight at all.

A small earthenware bowl was brought out of the hut, containing a strange concoction which I learned was bullrush fluff soaked in paraffin, the fuel of the flare. A little lump of it was lit from the torch and placed on an earthen platform on the very bows of one of the punts beneath the hood. This hood, made of reed matting, was about two feet high, built up behind and over the flare to make a shadow, and the *murdabchi*, armed with the net, took his place standing immediately behind it. A place was made for me to sit in the middle of the boat and one of the others climbed into the stern with a pole and pushed off along the narrow ditch. Closely following came the second boat, poled curiously enough from the bows, with the gong-beater in the stern.

Almost as soon as we left the hut the gonging began, at first softly, then gradually rising to a high crashing ring which drowned all other sounds of our progress. For the next half-hour the gong rang thus, rising and falling in intensity by reason more, it seemed, of its strange harmonics than of any intended variation on the part of the ringer.

The punts drove forward at a good speed along a narrow, winding path cut through the reeds and bushes. A mallard rose about five yards away, the splash faintly audible above the ring of the gong; it hung for a moment looking pale yellow in the light of the flare and then disappeared into the darkness.

Suddenly we came upon a pair of mallards sitting in the water half under a willow bush. The boat swung towards them as they swam out into the open. For a moment they paused and then rose. The net came sweeping across and the drake was taken six feet in the air. The net was held upright and the bird fell down into the pocket near the handle whence the catcher removed him, locked his wings and dropped him into the boat. The whole performance was accomplished with a speed and perfection which could only have been born of long practice. The punts never stopped or even slowed up, but swept on along the narrow waterway, propelled with absolute precision round sharp bends, under overhanging branches and through narrow gaps in the great reed beds. A teal jumped from thick rushes on one side and was caught right at the top of the net; another mallard splashed up from a clump of dead docks and was caught by a lightning stroke. A gadwall swam out from the reeds straight ahead, so close that as he rose he was taken in the pocket of the net right against the handle. Many would jump just out of range of the net. The catcher knew exactly when they could not be reached and stood perfectly still, making no attempt unless he were sure of a catch.

Sometimes teal and even the larger ducks would swim away low in the water, trying to hide. If they came out into an open place it was possible, by a special stroke of the net, to scoop them out of the water, but there was a much better method. The catcher would squeak by sucking through his lips, as if he were calling to a dog, and the sound was apparently audible to the duck above the din of the gong. This would make the most unwilling duck rise at once, only to be swept up in the net as soon as it was clear of the rushes.

We wove our way backwards and forwards through the marsh, along paths which were often, I afterwards discovered, parallel to each other and no more than ten yards apart. So little were the ducks affected by the passing of the boats that those ten yards away were left quite undisturbed. One part of the marsh seemed empty of ducks, and looking up into the bushes which in many places met overhead, I saw that the branches were full of dwarf cormorants. They were neat and handsome little birds with long tails, which they sometimes spread in a fan to balance themselves as they looked down into the glare of the light passing below.

When we emerged from the cormorant bushes the ducks seemed a little wilder than before. The *murdabchi* turned to me and pointed upwards. Above there was a great clear starlit patch in the sky. He shook his head and said 'Khoub neest', which means 'No good', and we headed back towards the hut. An overcast sky and complete darkness, I learned, is one of the essentials of duck-catching; rain is better still and snow is best of all.

During the half-hour we had been out just over twenty ducks had been caught. In the hut the wood fire was revived from its embers and we sat round waiting for the sky to cloud over so that we might set forth again. But after an hour the starlight was brighter than ever, so fitting ourselves like a jig-saw puzzle into the available part of the ten-foot-square floor (for the middle was occupied by the fire), the five of us lay down for the night. It was *not* a comfortable night so far as I was concerned, and when Mahmoud the gong-beater discovered that my leg made an excellent pillow, it became less comfortable still. But I occupied my time in thinking of the extraordinary things which I had seen and of their application to the catching of live ducks in other parts of the world.

I was determined to try and catch some myself and also

to learn as much about the method and the significance of its various features as I possibly could. So I decided to stay on at Siah Derveshan.

—

There followed several nights of clear starlight, and it was not possible to go out in the *murdab.* I spent the day with my clap-net trying to catch the greylag geese down on the 'pig-puddles'. They were the eastern pink-billed race, and quite different from our British greylag. But without decoys it was an impossible task. The geese were feeding over too large an area, and the 'pig-puddles' were too wet and bare to hide the sixteen yards of net successfully.

I spent four whole days in the attempt. On the first the net was set near the back of the bay, and I was able to lead the release wire to a tree at the edge of the wood. I climbed this tree and stood motionless in the shadow of the trunk, on a limb twenty feet from the ground. I stayed there from dawn until about midday. A white-tailed eagle settled once in the branches ten feet above me. I talked to him, but at first he could not locate the sound and was not at all alarmed by it. After several minutes, however, at last he saw me and flapped off in a great hurry. The bushes near by were full of bramblings, shining brilliant orange in the morning sun. But the geese did not come near the net. They fed over on the far side of the bay, and in the afternoon Ayub and I moved the net. Ayub was the nephew of the *mur-dabchi*, a tall, handsome youth, and inordinately proud of the rubber waders which I had lent him. We conversed, chiefly by signs, and in the few key words which I knew, like 'here' and 'there' and 'this' and 'that', and we had the net set out in record time, ready for another attempt in the morning.

The next three days, like the first, were cloudless, and the sky was brilliant blue. But the weather was cool even at midday, and very cold before the sun came up.

I had built myself a tiny hide of dead rushes and had roofed it in. To it ran the release wire of the net, and before dawn each day I would take up my position, and have my feet covered over with sticks and rushes by Ayub, who then retired to the edge of the wood. Each day I lay there from dawn until late in the afternoon, and arose stiff and disappointed but still hopeful of the next morning, when some new plan could be tried.

Walking out in the stillness before dawn the call of the waking Bewick's swans, who always roosted off the eastern point, was an additional help in direction finding; and when we were setting the net in the first grey light we sometimes saw an old boar go splashing back through the puddles towards the wood.

I would just have time to make sure by the pale daylight that the net was in order, take up my place in the hide and be covered up by Ayub, before the pintails and mallards came out from the *murdabs* inland. In great packs they would come tumbling from the sky to settle all round me on the pig-puddles. A little later the geese would come in from the lagoon, just as the snow mountains glowed with the unbelievable pink of the sunrise.

Throughout each day these greylags and whitefronts and mallards and pintails fed round my hide, sometimes within five yards of me, yet never would the greylags, which were my quarry, feed near enough to my net to be caught. Each day I tried a different method of concealing the net, and each day I was full of hope. At last one single bird stood within reach, or so it seemed, and I tugged at the trap-wire. But the pintails had dibbled about amongst the net so much, earlier in the day, that they had tangled it up. The net swung over slowly in soggy hanks instead of spread-out like gossamer, and the astonished greylag had plenty of time to fly out and make off with the rest of the army of wildfowl which had, so shortly before, been feeding peacefully around.

So ended my netting attempts on the pig-puddles, and that night a storm blew up. The first of it was a strong warm wind out of a clear sky, and as I left the bay a great column of soaring pelicans swept along the shore. They were rising in a gigantic spiral without flapping their wings at all. As they swung round and round, some just starting from the lagoon, others already no more than specks in the sky, they shone brilliantly white in the afternoon sun, by reason of the pinkness of their plumage. Long after I had left the shore I could see them, a hundred strong, curling upwards on the thermal current which they had found, and which took them to so great a height without effort.

When darkness fell the storm clouds blew up, and it began to rain. At last I was able to go again to the *murdab*, and try my hand at wielding the duck net.

As before, we met in the little hut beside the log fire, where the smoke soon reduced me to tears. But after half an hour the gong was tied on to its forked stick, and we sallied forth into a thin drizzle. The *murdabchi* came to instruct me, and stood behind me as we pushed out along the narrow waterways. The scene was lit by the flickering flare, and the din of the gong filled our ears.

I stood, at first rather insecurely, behind the hood, holding the net ready. The boat was very tipply, and I did not feel very confident that I could wield the net at all and still retain my foothold. The eerie light, the fantastic noise, the thrill of coming, at any moment, upon a duck that would jump up just in front, produced such a feeling of suspense and excitement as I have seldom experienced. The whole thing was so strange and improbable, and yet here was I actually catching ducks myself in this unbelievable way. The ducks were very numerous that night, and almost as soon as we had started my attention was fully occupied. Teal slipped out from amongst the reeds, mallards jumped up with alarming suddenness. To co-ordinate the wielding of the net with the unexpected appearance of a duck looking bright and clear-cut against the dim background of water and marsh, needed a high degree of concentration.

The hardest thing, I found, was to know when the birds that rose were really within reach. If they were not, and one made an unsuccessful stroke with the net on one side, that was always the moment chosen by a teal to rise quite close on the other. By the time one's balance was regained the teal was already out of reach.

I missed a great many ducks that I should have caught, but I managed to catch a few. One mallard duck swam against the side of the boat, and I was able to bend down and pick her up by hand, though I nearly capsized the boat in doing so. I scooped a white-eyed pochard out of the water, not perhaps with such a stylish stroke as it might have been, but none the less a successful one. I caught teal and gadwall and mallard and coot, and for every one that I caught I missed four that the *murdabchi* would have caught.

The drizzle seemed to make the ducks very loath to jump, and the teal had to be squeaked at in order to make them rise at all. Once I could have picked up a dabchick just in front of the boat, and later I tried to pick up another mallard duck by hand. She splashed so much in flapping along the surface of the water, however, that she extinguished the flare, and as soon as it was out, got up and flew away quacking loudly.

The boats were of primitive construction, and during the evening they leaked. It was practice, therefore, to stop occasionally to bale them out, and when we did so the gong-beater would usually take a short rest too. The instant that the gong stopped ringing nearby ducks began to rise, and in a few moments every one within fifty yards would be gone.

I noticed this, too, on later occasions, and I have no doubt that the gong plays an even more important part than I at first supposed. It is difficult to locate the noise, and almost impossible to judge how far away it is. Several times I thought that the second punt must have been left far behind, but, when I looked round, there it was, following close upon our stern. There is some subtle baffling quality about the combination of sounds, the clashing as of cymbals, and the deep ringing as of bells, which is bewildering to the ear of bird and man alike.

At the end of an hour I had myself caught a dozen ducks,

and then I watched the *murdabchi* at work and learned much from watching.

On the following night again I went to the *murdab*, and since it was still raining, conditions were favourable to the catcher. In an hour he had netted sixty-five ducks, including shovelers, pintails, gadwall, teal, and mallard. For February this was a fair bag, but I heard lurid tales of the dreadful destruction wrought amongst the ducks in the autumn when they first arrive from the north. My friend had that very season caught six hundred in one night, and across the river the *murdabchi* in Khinde Khale had caught over a thousand.

Yet the ducks seem able to cope with this drain upon their numbers, for I was told that no noticeable decrease in the duck population other than small annual fluctuations had been observed within the memory of the local duck-catchers. Round the shores of the lagoon I myself saw such quantities of ducks as made it impossible to believe there could ever be a shortage. In one bay I estimated that forty-five thousand teal sat roosting on the soft mud. Elsewhere thick clouds of mallards rose from reedy marshes, and convinced me that the winter population of the lagoon was to be measured in millions rather than thousands.

Yet one must beware of being led astray by such mass concentrations of waterfowl, because for every one such centre that exists today there may well have been a dozen in past times. Ornithologists are agreed that wildfowl in North America have been more than decimated in the last century, and yet it is possible nowadays to see as many ducks in the air at once over one of the well preserved 'refuges' as it was probably ever possible to see anywhere, even in olden times.

Wildfowl *throughout the world* are in nearly, if not quite, as great need of the far-reaching and far-sighted protection which they are now enjoying in the United States and in Canada.

I was sorry to leave Siah Derveshan, where the catching of ducks was not solely a means of livelihood to my very hospitable host, but a highly skilled art in which he delighted to excel. In the ten days that I had spent there I had learned much of Persian rural life; I had learned the intricacies of a fish trap and how to take 6-lb whitefish in it; I had learned that the art of throwing a cast net was not to be mastered in a day; and most interesting of all, I had learned about the strange light which moves eerily through the Caspian swamps at night, lighting the reeds and branches with a mysterious glow, like Edward Lear's immortal 'Dong with the Luminous Nose'.

From *Wild Chorus*

16

Conscience

Man is the most dangerous and destructive predatory animal the world has ever known. He is also the most imaginative and creative animal. Man has evolved important differences from other animals, like reasoning power and accumulated experience which are (arrogance or no arrogance) rather striking. Not least among these differences is the evolution of a conscience.

It is this conscience which suggests that wiping other species from the face of the earth unnecessarily and at least four times as rapidly as in prehistoric times may not be in the main stream of human evolutionary progress.

From a letter to the *Geographical Magazine*

17

The Guides Return

Sunday 8 July. A lovely hot morning. We all sunbathed outside our mess tent. For the first time the flies have begun to be really troublesome. It's not that they bite – much. An occasional one does; but the main trouble is getting them in one's eyes, mouth, hair and ears. They are a bore, but still only a minor bore. There are a good many species – all small dark Diptera. Some are quite black, others the same as the ones at Mývatn, with grey bands on their legs (*Simulium vittatum*). There are also blowflies of a different kind from the English ones, and dung-flies which look just the same as their English relatives. There are beetles with a green thorax and brown wing-cases, and other round beetles which are small and black. One species of moth – possibly a *Plusia* – is fairly common, and there are at least two other Noctuids and more than one species of Geometer.

After an idle morning, lying in the sun and writing up notes (how difficult* this is with so congenial a party, with so much pleasant talk and constructive discussion), and an early (for us) lunch, Phil and I set off to photograph the baby red-throated

*We counted ourselves lucky to have kept up to date with our diaries and notes during this trip. In most scientific field expeditions note-book time exceeds actual field-time – and if the time spent writing up data and conclusions for publication, on return home, be also included, the field-naturalist almost invariably finds that for every hour of observation he has mortgaged ten hours with pen, brush, or slide-rule.

divers. We got held up on the way photographing flowers. When we reached the Diver Lake only one young bird and one parent were to be seen on the water. I think the adult was the female. As I was filming her from the bank at fifty yards' range I heard a strange harsh squeaking coming from the far side. When I went round I found at once the other baby diver. It was sitting on the shore about ten yards away from the nest. It raised its bill as if to beg for food when I approached. I filmed it and then went on to film the adult with the other baby in full sunlight at about forty yards. This should be quite good. Phil photographed the baby on the shore, and then I caught it and we tagged it, filmed it in the hand to show the way the legs come out of the very hind end of the body, so it seems. The down was very dark blue-grey, bill, legs and eyes black.

We moved away up the hill beyond the lake so as to let the diver back to its second baby as soon as possible. We saw it climbing ashore when we were three or four hundred yards away.

As we came into sight of the upper pools of Chipping Swanmarsh (the swan's nest had contained chipping eggs on the day we had first visited it) we saw a pair of swans with one cygnet. This was quite evidently one we had previously called Cedric. They made off down the marsh. At one point they had to run the gauntlet of an angry Arctic tern through whose nest territory they were passing. The tern dived on them time and again, and at every dive the adult swans ducked. It was an astonishing performance, and quite obviously the swans didn't like it at all. When they finally got through they started overland, trailing little Cedric, to the next set of ponds in Chipping Swanmarsh.

We went over the ridge and surprised a pair of pinkfeet. They flew round and round and evidently had a brood or perhaps a nest. Their behaviour indicated that they had been disturbed before and we did not go hunting for their goslings because we thought it quite likely that they were the recaps of the evening before.

The alarm notes of this pair of geese must have been heard by the swans of nest T [9] because, when we topped the ridge, both birds were far out in the marsh. I filmed them as they flew up the valley. We went to the nest, and found that the eggs were not covered, which was rather remarkable since the bird had left the nest on a pinkfoot's alarm-note. We wondered whether perhaps some swans at least do not cover their eggs, or whether her departure had been more hurried than we supposed. While photographing the nest the female came back looking magnificent against the brilliant light shining down over the Kerlingarfjöll, and reflected from the pools in the marsh. She settled about 100 yards away and whooped in melancholy fashion every ten seconds until we had disappeared back over the ridge.

For some time there had been heavy rainstorms hanging about; indeed there was so little wind that hanging about is just what they did. One such had been hanging over Nauthagi (a place in the Upper Illaver) and Upper Oddkelsver for an hour or more. Now it extended, more than moved, towards us. There was a brilliant rainbow, for the sun continued to shine in under the raincloud. It did not rain for long, because suddenly the cloud above had spent itself and dissolved and vanished over us. Once more the sky was blue overhead and the flowers glistened, and it was warm again. We came slowly home to camp.

In the evening we made some kind of contact with Gufunes Radio. 'You are very weak,' he kept saying. 'Will you give me a long tuning-call?' And so James and Finnur took it in turn to wind the handle very fast – a most exhausting affair for them. At length he signed off: 'Understand you have no message for us. We have no message for you, will call again at 23.00 tomorrow night.'

We still have not passed the position of our camp, which we are anxious to do.

—

Monday 9 July. Just such another day as yesterday. We were having breakfast in the sun just outside the mess tent when I looked up and saw, standing not more than 100 yards away and looking at us, a ewe and a well-grown lamb. The sheep had long wool and horns. These were the first of what Finnur thinks will be a big influx of sheep into this area during the next week or two. The two sheep trotted on and we watched and filmed them as they went down to the Blautakvísl and eventually swam across. The ewe was evidently bound for some special place she had in mind.

Later in the morning I went down to the *kofi*, where James and Finnur had put up the hide in front of the snow-buntings' nest in the wall. There are two entrances to the nest (which apparently hatched about 29 June). The right-hand one goes out towards the river, the left-hand one on the downstream side of the building next to the door of the *kofi* itself. This *kofi* is a house made of stones, with a timber roof covered with earth. There is quite a rich flora growing on walls and roof. The *kofi* has a door but no windows. It consists of a single chamber, with a raised shelf across the dark end on which, presumably, the shepherds sleep. The shelf is occupied by saddles and other equipment for the horses at present, and a shelf along the back which is a kind of manger has Finnur's specimens on it.

I put some saddles in the hide to sit on and installed myself for three hours, during which it became so hot at one period that I had to take my jersey and shirt off, and so cold a little later that I was quite perished.

The snow-buntings came and went. I filmed them and made notes. The most interesting points were that the female always entered by the right-hand entrance and left by the left-hand one, whereas the male only used the left-hand one for coming and going. The female came more often than the male. Sometimes she brooded the young. The excretions of the young were brought out and not eaten, so far as could be ascertained. Most of the food brought was black and probably entirely insect food, but the female brought one feed which was entirely green. (Later we found that this was insect food too – sawfly caterpillars which both male and female snow-buntings found quickly on the glaucous willows.)

At twelve minutes past three, when I had been almost three and a half hours in the hide, James came down to see how it was going. The female snow-bunting had been sitting for twenty minutes. She flushed off, and we moved the hide in closer before returning to camp for a belated lunch.

Now I am sitting in the tent and am alternately too hot and too cold. This morning's showers have persisted, and it is raining now, with a cold draught coming in at the door. Not many minutes ago it was almost too hot to stay in the tent. There was a pattering on the tent roof then, but it was from inside instead of out. Several hundred assorted flies have come into the tent and they show a marked positive phototropism, that is to say they follow the sun round, crowding into that corner of the tent roof nearest to the sun.

But when it starts to rain their pattering ceases, and now the loud rattle of the rain drums on the tent roof. James has just returned drenched from his measurement of the home goose-nesting area. Finnur is in the hide watching the snow-bunting, and I am up to date with this journal for the first time for DAYS. Hurrah!

We are expecting the ponies with the second lot of our food tomorrow – 10 July. That is the day they promised not to be later than. Of our twelve army food-packs we have eaten seven. These seven boxes represent 70 rations, and we have eaten them at the rate of six people for the first three days (18 rations) and four people for thirteen days (52 rations), total 70 rations.

The ponies should bring a further 15 boxes, which will give us 20 boxes = 200 rations. One guide will be staying with us, so that if we are still here until 9 August, i.e. another 30 days, we shall require 30 by 5 = 150 rations. This will leave us with 5 boxes over at the end, out of which the return journey of three days at 6 or perhaps 7 people will take a further 2 boxes. We are likely, therefore, to have carted 3 ration boxes too many up here. Since, however, we did not know at what speed we should go through them I consider this a perfectly prudent error.

Very heavy rain this evening – much talk of whether the puffin or Wilson's petrel is the most numerous bird in the world. I persuaded Finnur to tell about the eiders again. He told us about a colony in Dýrafjördhur at a farm called Mýrar, which is a new colony started during this century and built up by the farmer Gísli Vagusson (or perhaps more, Finnur thinks, by his wife). It now gives 50 kilos of down per year and must contain more than three thousand pairs. Dýrafjördhur is on the west side of the North-West Peninsula.

Finnur thinks there are four hundred thousand eiders' nests in Iceland. He told me again about his time on the island in Hrútafjördhur, which was on a farm which belonged at that time to him. His grandfather actually built up an island for his eiders, which subsequently held two to three hundred pairs.

One delightful piece of eider behaviour which he described was a variant on the egg-covering habit. If the eider female was unable to cover her own eggs because the intruder advanced too fast, she would stop a little farther on and cover someone else's nest. This is a perfect example of an innate action serving a social purpose for which it can hardly have been evolved in the first place.

In the evening we had the best radio contact yet. We ascertained that there were no messages for us and passed the position of our camp – 'at Bólstaður on the west side of the River Thjórsá'. We asked him to pass this position to the British Minister. We felt well pleased, and I think James and Finnur felt reasonably well rewarded for all their efforts.

Tuesday 10 July. Another bright morning. I went into the marsh to catch Apus (*Lepidurus arcticus*), the primitive crustacean which looks like a tiny king-crab. One pool about four to six inches deep had a much greater number than any other, though there were a few of the creatures in one or two other pools. It seems that these pools must dry up in the later summer, but of course Apus can survive this.

I caught about fifty or sixty and filmed them in a bowl, together with a caddis-fly larva and some pond-snails (*Limnea ovata*). Afterwards I preserved the Apus in methylated spirit for Harold Munro Fox. Phil went down to the snow-bunting hide, now at eight feet from the nest-hole, and later we had a picnic lunch while sunbathing. We had thought of a plan for setting the net for half a dozen broods of pinkfeet which we could see through the heat haze on Oddkelsver, but by lunch time they had moved down on to the sand and were washing in a pool. This spoiled the plan, for we had intended to creep along the sand and set the net below the brow, then drive them down off the marsh, with flankers to direct them.

In the afternoon the geese were still at the pool on the sand. I decided to go over the Blautakvísl in order to visit the swans' nest by which we had temporarily put the hide. From camp I had seen that she was off the nest. It was so warm and sunny when I left that I wore my rucksack over a bare top half.

The geese saw me coming soon after I had crossed the river, and immediately started to go to the right towards the Thjórsá. I kept out as far as I could in order to head them off and went down quite suddenly in a quicksand, or rather quick gravel. It had a hard bottom about two feet down but it was unpleasant because so sudden. The interesting thing was that the geese seemed to recognise the security of the sands and the river, for they were not hurrying unduly. There were sixteen adults and either seven or eight broods. The important part was that a number of the parents were flightless. This is the first definite evidence we have had of flightless *parents*. The goslings with them must have been between two and three weeks old. Thus, supposing them to have hatched about 20 June, the first flightless non-breeders were seen about 3 July and the first flightless parents on 10 July.

I climbed up the brow and could see, across the marsh, the swan sitting on her nest. As she was still sitting I decided not to disturb her, especially as a rain storm was creeping up the river. The sun had gone in, so I turned back. Two broods of geese which were up on the slope beyond the swan – about half a mile away – started to move off up the hill.

On the way back I noticed a ringed plover running along the sand in front of me. He looked quite pretty against a pool, and I wondered if I should photograph him. I had just decided not to when he began to bathe in the pool, about fifteen yards away. I got out the camera and set it up on the tripod. The plover still bathed. I got out the big lens and fitted it to the camera. The plover had finished bathing but was preening charmingly. I estimated the range and set it on the lens, I got out the exposure meter and ascertained the light. The plover was preening away and looked delightful. I set the stop and put my eye to the view-finder. The plover flew away. So I took off the big lens, put the camera away and started on homeward again. Later I found a golden plover with a flock of about twenty-five dunlins in attendance – the largest flock I have seen so far. In such numbers it seems that the plover might be quite embarrassed.

By this time a strong south wind had set in which was very cold. I put on all the clothes I had with me in my rucksack. The cold wind had completely banished the flies which had been troublesome early (although still only a small proportion of them bite). On the way back to camp I noticed an Opilionid (harvestman) carrying a small fly in its jaws. It was the long-legged species with the black mark down its back (*Mitopus morio*). The impending storm was behaving in a very strange way, for in England it would have been upon me in a few minutes driven by the stiff breeze. But here we have to get used to the idea of clouds forming and dispersing at certain given distances from the glaciers – or, at least, I suppose the glaciers to have some such influence. Anyway, the storm crept up the Thjórsá so slowly that, although we were equidistant from the camp when I turned back, I beat it home. Within two minutes of my return a great downpour came upon us. But our tents were waterproof and we were perfectly snug within. After the rain had passed – or stopped anyway – we began to discuss whether the ponies would arrive today. Finnur said that originally they had thought to *start* up on the 10th. I thought perhaps they would stick to that plan and perhaps be delayed a day, and maybe we should not expect them until 14th or 15th. At that precise moment we heard a sound. Swans, maybe. Finnur bobbed out of the tent. 'The horses are here' – and so they were. Three guides came, one of them Valentínus (Valli) Jónsson from Skaftholt, one called Ágúst Sveinsson from Ásar (the district postmaster) and one a fox-catcher Fílippus Jónsson of Háholt. There were ten ponies in two trains. Ten minutes later they were being unloaded in a welter of ropes and straps and pack saddles.

It was a great event. We filmed and photographed it, and then we went into the tent to brew tea and to open the packages of mail and newspapers. Newspapers were the bulk of it, and a fat interesting-looking envelope on His Britannic Majesty's Service from the Minister contained only two lurid novelettes. There was nothing of interest from home, no news of the breeding successes, no news of the ducks from Mývatn – nothing but a tiresome detail for my Perry River book, and a boring letter from my lawyer. It is extraordinary how such a small thing can be so disappointing. Phil and I were quite crestfallen about it. James had several letters from his wife and Finnur one from his.

The newspapers brought the world into our camp and the change was not for the better. True, there was to be an armistice in Korea, but the Persian oil dispute is not settled, We are in no way the better for being reminded of all these things and the periodicals threaten to waste a good deal of our precious time up here.

In the evening we had a really successful radio communication with Gufunes, better even than last night. He heard us, strength two. We told him we had no messages for him but that our relief supplies had arrived by pony. (He couldn't get the word 'relief'.) I gave him 'food supplies', and he said: 'I understand your meat supplies, food supplies have arrived by pony.'

There were alarms and excursions because two of the ponies had started for home. Valli went after them and managed to get them back. He is to stay with us – with seven ponies – for the rest of the expedition.

From *A Thousand Geese*

18

The Flowers of the Eagle's Fell

Friday 20 July. Windless with high overcast. Rather a warm day, and the flies a nuisance. We spent the morning packing up our sleeping tents and night requirements for a three-day trip, and, as I write, Valli is saddling and loading the ponies. With James's new hand-nets we shall look more than ever like a band of White Knights. Finnur has been down to the *kofi* to get his plant-collecting equipment. He reports that the first baby snow-bunting has left the nest. The day improves as we set off – to work the nets at Oddkelsalda, and on to Arnarfell.

And a splendid day, too. At 1.30 we set out across the Blautakvísl and went straight towards the Foxwarren. As we came to the gap west of the Barnacle Marsh, Valli skilfully noticed geese ahead. There was a little ridge about 200 yards away. As we galloped over it there came into view a great bunch of geese – I thought about a hundred – and Valli and Phil were on top of them. I went out to the left and caught an adult and three young. Phil caught no less than five adults. James, who, because he has a slow pony, now takes charge of the pack-horses, caught three goslings which had broken back, and after just under an hour we had ringed and released our record catch – twenty-eight geese, of which thirteen were adults.

One of the main features of this success was the new hand-nets which James had made from the uprights of the photographic hides with a wire hoop and a loose piece of string netting. They were admirably useful.

The drive to the nets was conducted by an approach from one side only this time. Valli was to gallop round the point along the Miklakvísl bank and out across the marsh. In this plan much depended on where the geese were in the marsh, and, as it happened, they were well away from our hill and headed, as soon as they saw Valli, for the main mass of Oddkelsalda. A few turned and ran up the hill, and some went on to the lake. I saw some on the hill-top and went round the lake. I came upon an adult crouching under the bank. I tried to use the net, crept up, but the goose had lifted his head and flapped out, just as I lunged. I tripped, fell headlong on the bank at the very brink of the lake.

One goose I had seen against the outside of the net. Valli went up and caught it, and perhaps a second, and then withdrew again. On my way up the hill I startled a crouching adult, but it gained on my reluctant pony and therefore would not crouch.

When I reached the top the others had got the geese into bags. There were six adults, which ultimately turned out all to be males, and four goslings. One adult had approached the net from the wrong side and got tangled.

Although we had ten geese it was a sad disappointment. Had we done the drive as we did the last time we should surely have had sixty at least.

We lunched after the Miklakvísl crossing. Thirty-eight geese before lunch was good, even if lunch was at 5.00 p.m. The flies were very troublesome as it was windless and rather heavy. Thin high cloud had gradually obscured the sun. Rain seemed imminent but who could say how imminent?

It was not until we had crossed both the Fremri and Innri Mülakvísls that we came to more geese – this time two unaccompanied goslings on a lake, which were finally caught after Valli had plunged in and was wading above his knees. A single gosling was peeping out on the sand at the edge of the Thjórsá but it would not come to us.

Last time we had been at Vallarvatn I had left my nice spring pliers behind. We hunted for them now, and James found them – the clue being a piece of string from a bunch of rings which had been exhausted on the four goslings we caught at this spot. I was greatly delighted at the recovery of these pliers, which I secured to my belt by the piece of string in question. Part of the reason for my pleasure was that these pliers are the best for ringing and do the job much quicker than the ones I had been using since the loss; part also, no doubt, was the rectifying of the untidiness of losing them; but a large part also was related to the loss of a toy and came from long ago.

We repeated the route of our last visit to Arnarfellsalda, going round the eastern side. Once more we found the swan with the five cygnets, once more we set out to catch and tag them, once more as we approached them we sighted geese near the top of Arnarfellsalda, and once more we left the swans to go after the geese. This time it may have been a mistake, for although, after a chase all over the top and down the other side, we caught two adults and three goslings, yet I think this could still have been achieved after tagging the swans.

The goose chase led down past a lake with a brood of three longtails to the Innri Múlakvísl, and there Valli and Phil surprised the white-tailed eagle at a gosling which he had just killed and which was evidently one of the brood we had disturbed. The eagle had evidently caught the gosling in the water as it was soaking wet. Phil saw it rise out of the water. The bird flopped across to a mound on the other side of the river, where it sat, being mobbed by a whimbrel. We crossed the river, but when I dismounted in order to film him, the eagle flopped off, turned and flopped past us, to settle across the river behind us. We found and photographed the gosling – headless, but still quite warm. Valli led me on my horse back across the river for another effort to film the eagle, which was a complete failure. He flew at 150 yards, and when I put on my rucksack the winder of my watch broke off. Valli and I looked for it, but a hopeless search. We marked the place with a feather, but I don't suppose we shall ever have a chance to look for it.

We had seen about twenty unaccompanied goslings in the marsh ahead of us, and in due course we came up with them. Eight of them came towards us – to within about 100 yards – after which they started off away from us and we in pursuit. So began one of the most arduous and least rewarding hunts we have undertaken.

Early in the hunt six goslings crouched in a bunch. Phil and I crept to within two or three yards, then up they sprang and ran in all directions. Valli chased up four more, and for a while we were running in all directions over soft mossy bog with occasional harder mounds of tundra. Sometimes in these mounds there would be a sudden subsidence as the ice cracked and broke below the surface. The bog was so soft in places that poor Phil went right through and fell forward, getting soaked to the skin, and wetting her camera again (the other camera this time).

Without James's new nets we should never have caught any. As it was we caught three before the area was clear except for four in a lochan. Valli plunged into this and for half an hour waded about almost up to his middle. The goslings dived hopelessly and Valli, in that depth of water, was insufficiently mobile. At last he caught one, and in due course the other three crept ashore, to be caught, one by Finnur, one by Phil, and one by me.

By this time it was almost 9.00 p.m. and we had but a short stretch of marsh to cross to reach the outer moraine. Here we decided to make camp beside a mossbound spring, in the lee of the steep richly flowered bank.

As supper was being cooked by Phil, James announced that he had seen nine goslings running up the grey stony hills of the outer moraine. Finnur, Valli and I grabbed nets and joined him in pursuit. The goslings led us (me in gym shoes) over the stones, worn round and smooth by the glacier, to a lake beside which there were already about thirty more goslings with four or five adults. Pursuit over this rough going on foot was obviously fruitless. We divided up and I went off round the lake. Three goslings were left on the water, and one adult swimming low ahead of Valli. I climbed a steep soft hill with strange 'quick-gravel' banks, and from there watched James crouching and waiting for the goslings to emerge. In due course they came, peeping, and almost ran up to him. But when he finally gave chase they got back into the lake. Nothing daunted, James went in after them up to his knees and caught one gosling – our fifty-third goose of the day.

Before leaving the top of my hillock I saw about forty geese on a further lake, to the south. They had seen me and were swimming away. It is evident that there must be many flightless geese on these morainal lakes, but it will be impossible to ride after them, so they will be a dead loss to us.

I came down the hill, past the largest and most wonderful clump of yellow *Saxifraga hirculus*, and so to ring the little gosling and have an excellent beef stew, at our new and quite comfortable camp.

Saturday 21 July. It stopped raining about 8.00 a.m., and intermittent sun made the tent very hot. Now I sit in the sun after breakfast. Finnur has already found some exciting new flowers:

Habenaria viridis (*Coeloglossum viride*). Frog orchid.
Epilobium sp. (probably *lactiflorum*). White willow-herb.
Botrychium lunaria. Maidenhair type fern. (This we have found before.)

He also brought some very fine *Veronica fruticans* and *Pyrola minor* and yesterday he found a new *Carex*, as yet unidentified.

Today we go on to Arnarfellsbrekka, the rich vegetation at the foot of Arnarfell (the Eagles' Mountain). Higher on the hill behind Finnur hopes to find rarities, for it was evidently a nunatak in the Ice Age and has therefore been bare of ice for a very long time.

7.45 p.m. We are back at our camp on the outer moraine in Múlaver, after a very pleasant, if less than usually anserine day.

No sooner had we started, after I had filmed some flowers – the orchid and the eyrarros (*Chamaenerion latifolium*) – than the eagle came sweeping in from the marsh and settled on the moraine. From about 200 yards I tried to film him, but he flew and settled farther into the grey waste of stones. I stalked this time and got to within 100 yards. The bird was against the sky and I got a very good shot of him before he took off and flew on along the moraine.

There were a few geese feeding on the marsh, but an attempt to drive them towards the moraine totally failed. They just went farther out into the marsh. We rode along the edge of the outer moraine with soft lush green bog on our right. A young dunlin, only just able to fly, ran up a mound. He had a short-tailed look – and a short bill too. This is the first baby dunlin, we have seen. At some places the moraine mounds are 30 feet high, but as we proceeded they became lower. At one point there was a wide expanse of stony river bed. Apparently one year the Arnarfellskvísl ran through there and washed all the moraines away – outer and inner.

Later we came to the shingle beds of this year's Arnarfellskvísl, and to the stream itself, which was the swiftest water we have so far had to cross.

The grey stones of the dry river beds are greatly enriched by patches, in places almost beds, of brilliant purplish pink – the eyrarros. The effect with a shimmer of heat-haze is marvellous on these great expanses of grey shingle, otherwise only broken by small patches of yellow-green moss.

Arnarfell is a steep red mountain with green lower slopes, about 1,600 feet higher than the surrounding plateau. It lies in a cleft between two great tongues of glacier, with a half circle of lesser mountains at its back. As we drew near, it towered above us, and a wild gorge bordered with screes opened out to the south of it. The glacier tongue is held in by an ice-washed bastion with very steep cliffs which seemed a likely place for an Iceland falcon. But as we approached the main mountain Finnur spotted a bird of prey standing on a green ledge far, far up. We thought it was an Iceland falcon, perhaps even this year's fledgling, but later, when we had dismounted and were looking at the flowers along the foot of the mountain, the bird took off, and soared out from the slope. It was our friend the eagle again. As we came to the bank my horse stepped on to a most innocent-looking patch of moss-covered shingle and went straight down with all four feet. I jumped off quick, and he turned so that I led him out, none the worse. Such a thing in the middle of a river would have been much less pleasant.

The flowers on the lowest slope of the mountain, a few feet up from the flat gravel plain, were indeed striking. There was, for example, a great bed of purple-blue cranesbills (*Geranium sylvaticum*), with one freak plant among them – almost white with a tinge of lilac. Among them, to make the colour by contrast richer yet, were buttercups and *Potentilla verna*. All this was in great profusion and richness, and there was also a lush growth of *Archangelica*, although this had suffered largely from the passage of the first sheep which we could see ahead of us. It seems that they have a predilection for the plant and go from head to head eating off the great bulbous umbelliferous buds.

We walked along the flat at the foot of the slope, finding upon it, and upon the rather steep bank which rises only a few feet above it, a number of new plants and a most luxuriant growth of many old ones. *Veronica fruticans*, the superb speedwell, was in evidence together with *V. alpina*. There was a new pink plant, very small and varying in intensity of colour, *Sedum villosum*; rich thyme and many saxifrages – *S. caespitosa* (*groenlandica*), *S. cernua*, *S. nivalis*, *S. stellaris*, and a new white one, very delicate and pretty, *S. hypnoides*. But king of all the saxifrages was the marvellous yellow *S. hirculus*. Hitherto we have seen it growing in ones and twos – at most four or five flowers – all over the brown parts of the tundra (the lumps and ridges in the tundra bogs). Sometimes among stones we have seen it, evidently thriving with up to a dozen flowers. But now suddenly we realised that the great yellow expanse which we had seen in the mirage a mile away, and taken for the very yellow moss which grows round springs, was in fact a field of yellow saxifrage growing in unimagined profusion, against a cushion of dark green moss. The star-shaped flowers of this lovely plant can be varied in shape, for sometimes the petals are full and round-tipped, at others pointed and more delicate. Here were all shapes and sizes – even a plant with near-white flowers which Finnur collected. For a flower which we already regarded as one of the most lovely up here, it was immensely exciting to see it in such a mass.

Under the bank ran a mossy wet strip with occasional springs, and it was clear that a number of geese had been here recently. There were droppings and many moulted feathers. The geese had evidently been feeding on the yellow moss, which was pulled up in beakfuls. But only a part of it must have been eaten, the rest lying about untidily on the smooth moss sward.

Among the *Archangelica* we found a sheltered bank for lunch. We chewed some flower stems of the great plant, but Phil does not like the taste and James found them too strong. We ate our corned beef and cheese, crisp-bread and biscuits, with dates for afters, and for part of the time it rained heavily.

After lunch we planned to go up across the foot of a small and now detached glacier tongue to the vegetation patches on a slope beyond where Finnur hoped for rare plants. We walked along the foot of Arnarfell to the edge of the glacier. A common plant here was the sorrel-like *Oxyria digyna* with little round leaves. On the face of the mountain was the old mark of the lateral moraine – the 'high-ice' mark, as it were. The glacier had long since retreated from here, but the stones had trickled down to make a protective layer under which was still ice – fossil ice almost. At one point there had been a subsidence of this ice, leaving an open ice cave and smooth glissades, with little landslides coming down it now and then. The cave was very impressive, the roof consisting of ice about eight feet thick, dripping continuously. Inside was a muddy pool, but the walls were of clear green ice and on one side we could see the stones bedded in the ice.

Here there were recent mud slides, there were furrows where huge boulders as big as motor-cars had slithered down. The soil was soft and new – a landscape in the making. Farther up we followed a small stream to its source under the lip of the glacier itself. There we found a young purple sandpiper, just able to fly, and filmed him.

We climbed a little way on to the ice, which was uneven and pockmarked. In some places the very end of the glacier seemed to be pushing the mud and stones uphill on to the terminal moraines.

At this point Phil discovered she had lost her watch – the little gold one she has had for seventeen years. She decided to retrace her course in hope of finding it. James went back with her and I went on to support Finnur and his botany. But when I had trudged through the soft moraines for a quarter of a mile I came up with Finnur, whose progress was barred by a roaring stream of turgid water. There was nothing to do but film its egress from the glacier and then turn back. Finnur had picked up a piece of obsidian – a lump of black glass about 3 inches long and 1½ inches wide. It is used, he told me, in astronomical optics, and a piece weighing several tons was shipped to America last year. We walked back, I feeling somewhat awed by the elemental nature of these surroundings. I do not really like these lifeless moraines. Back on the green lower slope of Arnarfell we saw two wheatears. We had heard one before lunch. The two were evidently fledglings, and it seems probable that they were hatched here, so this is a doubtful addition to our list of breeding birds.

We followed the wheatears up the slope a little, and from there looked out across the expanse of shingle, intermittently green between the wide beds of the glacial stream. There was a group of about twenty geese, mostly adults, feeding out there, perhaps the geese of which we had seen signs along the foot of the mountain.

Back at the lunch place, Phil had not found her watch and James was asleep in the sun. We collected our bits and pieces and set off to ride back to our camp on the outer moraine of the main glacier tongue. On the way we started a female teal from a little open stream in the shingle. The crossing of the swift Arnarfellskvísl was negotiated without mishap, at the same point where the tracks of two lorries came to the river. These lorries, with about thirty men, had made an expedition up here last year, in late summer, in order to discover whether a tourist road could be opened in this part of the interior. They came up the east side of the Thjórsá, crossed it higher up and came back along this side. It seems unlikely that the country will be opened up in this way, and even if it were it would only be in late summer, when it would be unlikely to affect the geese.

We rode along the crest of the uneven outer moraine all the way home counting pinkfoot nests – hatched and empty. From the start of the moraine to the Innri Múlakvísl – that is to say the part of the moraine which is in Arnarfellsver – we counted twenty-six nests, although I had a private list of seven which *could* at a pinch have been nests of the previous year. Nests with little down can be very deceptive, and the one near camp which hatched the day after we arrived, but had no down, now looks about three years old.

After crossing the Innri Múlakvísl (not difficult where it comes through the moraine) we found nine more nests, plus three in the immediate vicinity of camp, so that up here there have been thirty-eight nests. We shall count from here on as we go.

Fifty yards short of camp we came upon a ptarmigan with small young. The young were about 4½ inches long, but almost unable to fly. We caught, eventually, five of them. The mother was very anxious and came fluttering round us, especially close to Finnur who had a baby in his hand. She ran about with open wings only three or four feet from him.

We tagged the five young birds and then came into camp, and a supper of stewed fish and puff-balls. It was only about 8.00 p.m. so we got to bed early. After supper Finnur went botanising and found two new plants, both very tiny: *Ranunculus pygmaeus* already over – a minute buttercup – and the other perhaps *Koenigia islandica*, an even more minute group of white flowers nestling in three tiny leaves.

During the night some geese came across the marsh near camp. At 5.30 a.m. I looked out, hearing continuous scolding from a whimbrel. I thought maybe it was the eagle, but I could see no eagle, only a crowd of fifty geese. The count was exactly fifty. There were about sixteen adults, the rest goslings. Two goslings were unattached but seemed to be getting along quite well.

I saw an admirable family row in which goslings of both sides joined in. There was much neck lowering and false preening in the adults, but no evident victor and vanquished as there usually is in winter. Both sides stood their ground and passion gradually became spent.

Sunday 22 July. It snowed, or rather sleeted early this morning. It's very cold. We plan to return to our base camp at Bólstaður by way of Nauthagi (pronounced Noitayi) where there is a specially large goose-fold. We must try to get some geese today.

Our day began with a delightful ride along the moraine, counting nests. The final total in Múlaver was thirty-three empty nests. A small number may have been of previous seasons, but not many.

The nest site of the pinkfoot is interesting. It is usually at the highest available point. This is particularly true in the marshes and tundra, rather less so on the mounds of the moraines. Here the gander's look-out post is at the highest point and the nest must be within about fifteen feet of it. There is evidently some important survival value in the two birds being close together, probably because only as a pair can they repel a predator – fox, falcon, gull, eagle, snowy owl.

It is interesting that we have seen all these in action except the owl. I do not think the Arctic skua is a first-hand predator except on unattended eggs.

A small lake just inside the outer moraine contained a pair of swans. They seemed rather tame. Finnur dismounted to photograph them against the background of the glacier at a range of thirty or forty yards, in bright sun. They looked splendid. One bird flew off, but the other remained and it soon became evident that it had moulted and was flightless. We surrounded the lake, but the bird had the sense to remain on the water – which was very deep in parts. So having ineffectually lobbed stones behind it to persuade it to go ashore, and having photographed it, we accepted defeat and rode on.

At one point on the moraine we came upon a bank of gentians. This was a small brilliant blue flower, *Gentiana nivalis.* It was a cold blue, contrasting with the warm blue of *Veronica fruticans* growing all round and among it. The yellow in the mixture was provided by *Hieracium* sp. – the hawkweed, now out in profusion on some parts of the moraine. At various points we had seen signs of the truck expedition of last summer – tracks and small cairns where they had camped for the night. Two such cairns together had been at Arnarfell. Now we found one with a bottle stuck into it. The bottle had a message inside. There was a poem, which Finnur declared was untranslatable, and there was record of the date (19 Sep. 1950) on which the camp had been made there.

In due course we left the perimeter of the glacier's fan and, crossing the Fremri Múlakvísl into Illaver, set out across a stretch of shingle, sparsely covered with vegetation. Ahead was a low wave with a prominent goose-fold on it. We skirted this at first in order to surprise any geese beyond.

We sighted one lot, far out, and galloped off after them, but they could go very fast on this kind of terrain. In some curious way they disappeared, but I saw some more running away to the right. The galloping cavalcade wheeled towards them, but once more they disappeared. This time, however, some of the goslings had crouched and we caught five of them – one with a lame left leg. We ringed it on the right leg, which is our convention for this year anyway.

Then we went back to look at the goose-fold. It was the best preserved of any we had seen, the walls still standing, and supported by vegetation growing in the sand which has silted up round it, inside and out. This fold is evidently contemporary with the first one we saw, near the summit of Nautalda, and I think, a good deal later than the much smaller and more ruinous ones we have found on the tops of so many hills.

But this one had two new features. First it was curved, almost like a decoy pipe, and secondly it was only about fifty yards from the edge of the flat shingle beds on the least promising side. Thus it is evident that the geese were driven over the hill from the other side and swept round into the fold, which led uphill almost back in the direction from which they had come.

The dimensions of this fold were about the same as those of the Nautalda one: 12 m. long and 2–2½ wide. At the blind end the wall was about 4½ft high. It seems possible that the wall was originally higher at the blind end than at the other, but this may be simply due to the drifting sands which have supported the wall at that end. There was a hatched empty nest against the wall. I am now absolutely satisfied that these folds cannot have been used without some kind of wings or leaders to funnel the geese into the trap. Had these wings been nets it is unlikely that a great stone structure would have been necessary. Why not a cage of net as well? Thus we must suppose that the wings were made of something which, although good enough for wings, was not good enough to hold a bulk of geese in the cage at the conclusion of the drive. For material for these wings we must look to the most easily available and cheapest. Horse-hair ropes, no doubt, were available, but probably fairly valuable, and a rope by itself although possibly effective far out on the wings, could not have been enough to steer the geese into the walled enclosure.

The easiest thing to get on the spot would be the roots and stems of *Salix glauca* and *Salix lanata*. These form a sort of tangle, with long roots which could be tied together and would form a network sufficiently impervious to geese for the wings of the trap but not for the cage. I believe the wings of these traps to have been made of this *Salix* scrub, anchored with stones.

The insides of the folds might well repay excavation. So too might a small square of stones which we found on the shingle flat below and which had evidently once been a house.

Twenty minutes later we had sighted another bunch of geese and we galloped down after them across a much richer meadow. The geese got to the Miklakvísl and we abandoned them, returning to the richest part of the meadow, where a small notice on a post proclaimed that we were at Nauthagi. Under the word 'Nauthagi' was the date 1928 and the letters OB, standing for Ólafur Bergsson, who put up the notice (celebrated by a big party with spirits and cakes) on the occasion of his fiftieth visit. He was farmer at Skíðufell just opposite Ásólfsstaðir. He was *fjäll kongur* ('mountain-king') that is, in complete control of all sheep-collecting operations, and is now buried at his farm, with his horse. Nauthagi means 'the Bull Meadow', and Nautalda means 'the Bull Wave'. It seems that a bull once spent the winter here by itself, feeding on the lush grass. We found out why the grass was so lush here. There were springs, and they were hot springs. In some of the creeks or rivulets, banked with mosses and rich vegetation, there were large numbers of sticklebacks. In one stream, where the water ran down a six-inch channel, there were a lot of dead sticklebacks which had evidently cooked themselves.

For the last two hours the wind blowing off the glacier had

been very strong and very cold. It was mid-wintery weather, after the snow of the early morning (which had left the Tungnafellsjökull and its perimeter of mountains all white). So we dabbled our hands in the warm water. We found one spring in which we could only just bear the temperature. It was most soothing.

But Finnur was having the most exciting time, for he had found almost a dozen new species of plants clustered in this little oasis of warmth. There were the dark red rather dried-up-looking flowers of *Comarum palustre*; the three-lobed leaves of bog-bean (*Menyanthes trifoliata*); there was yellow rattle (*Rhinanthus minor*); and a tiny delicate white *Stellaria crassifolia*; a new willow-herb (*Epilobium palustre*), to me very much like *alsinifolium*, there was the maidenhair-like fern *Botrychium lunaria*; and most exciting of all, its very rare relation *Ophioglossum vulgatum*, a smooth tongue-like leaf with a green flower still in bud, not more than three inches high (snake's tongue), unexciting to look at, perhaps, but so rare, Finnur told us, that its localities in Iceland are each mentioned by name in the *Flora*. This is a new locality for it. There was also a pretty moss holding silver drops of water on it, pale grey-green and unusual.

While Finnur was botanising, Valli had somewhere found a gosling, which was duly ringed and released. I collected some algae from the hot springs for Edward Hindle. We did not have a thermometer with us, and he had indicated that the exact temperature was important. But there seemed to be a fairly wide latitude in the requirements of this particular alga, between water which was only just bearable to the immersed hand, down to water which just felt warm – i.e. from 110°F down to about 75°F. James collected a spider, which was probably the same as we have found already.

In the middle of the cluster of springs (some, incidentally, cold) was a mound with a goose nest on it, and *Sedum roseum* growing on it.

This was evidently a traditional nest of great antiquity, and it is exciting to think of the generations of geese, perhaps reaching back to the days of the goose-folds, seven centuries or more ago, which have been hatched at that precise spot. We found the flank feather of a drake teal among the hot springs, and then we rode on, crossing the Miklakvísl and coming to the cold springs at the foot of Nautalda, where we had lunched before. Here we lunched again, in a cold wind with a huge dark cloud hanging above us. This cloud, in the lee of the glacier, was quite stationary, a front with blue sky over the ice and cloudy greyness perpetually over us, so it seemed. Two new plants here, *Juncus arcticus*, a very ordinary-looking rush, and *Selaginella selaginoides*, the familiar lycopod, just now beginning to appear.

I put on more clothes from my rucksack and felt warmer as we rode on round the foot of the hill on the north-west side. We passed a single goose walking out on to the marsh. Perhaps because it was alone it walked out in quite a leisurely fashion. But the stream between us and it was reputedly too soft to follow. At the south end of the hill we came upon about forty geese and a long hunt ensued. The geese went out across a wide shingle stretch beyond the hill. Unfortunately a soft patch of marsh held us up, but Valli and Phil got through this. Had the shingle beyond been hard we might have made a big catch, but it was soft enough to prevent the ponies from going really fast. James and I tried to go round the marsh, which took us far off our course. Finnur came on behind with the pack ponies.

When we got out on to the shingle we found many soft places which looked unbelievably innocent. James got into one of these, and his pony went in up to the hocks. He just got it out without having to dismount. I went off away to the left in pursuit of two goslings, and eventually Phil came too, and we got them both.

We gathered together on this open tract of gravel for the ringing operation. The wind was now gale force and the battle of the clouds was at its height. An enormous sharp-edged cumulus cloud gave space to the sun, and we were at once warmer. We had caught two adult geese and five goslings, but two of them were almost fully fledged. These must have been at least six weeks old and were taken for adults during catching.

As we rode on towards the Blautakvísl the weather once more became a feature of our day. The wind came in gusts and whirls which brought clouds of dust, sweeping up in ominous yellow smoke wreaths. We got caught in it for a moment and our horses began to trot, whipped into it by the cutting grit. And then it had passed and we watched the yellow cloud blowing away. The wind was still chill, but the day had suddenly smiled on us. We rode across Blautakvísl and on to Ptarmigan Lake, but no geese. Over the ridge beside an attractive lake, which we have called Philsvatn, James cried 'tally-ho' and a new chase was on, rather a long one, as the geese were about 500 yards away at the start. Eventually, however, Phil and Valli got amongst them, on a low stony ridge (on which was yet another small early goose-fold) and the birds crouched all around.

This was very typical of our hunts and emphasised the highly sporting character of our ringing activities. A sharp gallop over awkward ground, after which some of the adults and most of the young crouch. But there is a marked distinction between the behaviour of the two, for whereas the adults usually remain crouched during a stealthy approach from the dismounting place ten yards away, and can be picked up by hand, the goslings always start running again at about three or four yards. The goslings can be caught after an all-out zigzag run of about 100 yards especially if they can be turned to run downhill, when human legs have the advantage. James's nets, large mesh net stretched on soft wire at the end of one of the aluminium uprights of one of Eric Hosking's hides (those of the other support our wireless aerial at the tent end) . . . these nets greatly facilitate capture at the last moment, and shorten the sprint by a most welcome amount. If the adults start to run again it is almost useless to pursue on foot, again unless they are going downhill. On this particular occasion, having twice failed, on dismounting, to pick up one adult, I did finally over-run it downhill and persuade it to crouch until captured. We forgathered with some of the catch on the summit of the ridge and then sighted another young one trying to creep away. Then two, and when Valli and I reached the spot there were three. Finnur came to the rescue and we caught all three. We had begun the ringing when Finnur suddenly spotted another adult, about 100 yards away, crouching by a big rock. It had shifted its position, which movement drew his eye to it. Again the hunt was on. Valli leaped to horse. The bird broke and ran for it, and Valli had a long hunt, ultimately driving it to James, who got it with a tennis shot. Thus at the end of a most sportive chase we had fifteen geese (four adults). After the hunt we saw two goslings making off in the opposite direction to that in which the ringed ones had been released. Valli went off after them, but found that the bird he caught was in fact one just ringed. However, the pursuit led Phil to a sandy flat on which she found a newly hatched ringed plover with an injury-feigning mother.

We made a detour to approach the hill lake above Pintail Marsh and then found that a small bunch of geese were on the far side, where we might have got them had we gone directly to it.

Our next geese appeared close in front of us at the very softest part of the middle of Deserted Swan Marsh. This was very bad luck, because the geese were so close at the outset that many were crouching around, and even *in* the muddy stream which drains the marsh. Valli and I went across and on up the hill over into the little valley which James calls Failure Creek, the valley next to Bólstaður and our camp. Here I caught an adult and Phil later had a great chase after a gosling on a lake, which swam from one side to the other. This was a most important gosling because it was tagged C. 506 – that is to say, it was one of the first two broods to be tagged and was from nest 4 or 5 (I did not write down the two separately). The most notable feature of this gosling is that its age was known exactly (twenty-five days), and that its size was strikingly small. It was only just unnecessary to flatten the ring. By this measurement of size it seems likely that my estimate of five to six days old for the young we caught in Oddkelsver on 2 July (which were a very average size when recaptured on 17 July) was too low. They may have originally been ten days old. If so, it must take the average hatching date back from 26 June to 22, perhaps 20, which in turn means complete clutches by 24 May and first eggs 12 May.

James had a fruitless hunt on foot right up the marsh, which left the pack ponies unattended, much to Valli's alarm and despondency, because the two got separated, and he fears a disaster with the ponies getting frightened and bolting, scattering the contents of the packs all around. James's excursion, however, disclosed a scaup with four small ducklings. This might be the bird from the nest we found in Chipping Swanmarsh, but is most probably a new one.

We returned to find everything all right at our base camp, and there were thirty-five geese in the hollow at the mouth of the iron stream, 100 yards from our mess tent. They went off down stream in the Thjórsá. Later, in the low evening sunlight, and amazing clarity of atmosphere, we watched about thirty geese on the sand at the mouth of the Blautakvísl, and more away across the Thjórsá sands at the edge of Thúfuver and Biskupsðúfa (Thoovoo-vair and Bisküpsthoova).

From *A Thousand Geese*

19

Slimbridge Discovered

In the autumn of 1945 I received two letters from ornithologist friends which, taken in conjunction, were to have a very profound effect on my life. Both these letters were from farmers and both concerned wild geese. The first was from Howard Davis, an experienced observer of birds living near Bristol, who sent me a copy of a paper he had written on the great flock of white-fronted geese which has wintered on the Severn Estuary from immemorial times. If I could spare the time to come down, he wrote, he would like to show them to me. I remembered my brief visit there before the war, at a time when the main flock had just left on the spring migration. I had seen, as I recalled, a bunch of twenty or thirty and that was all, and I had examined with interest the old duck decoy in the little wood. It would be nice to go there again, but wondered when, if ever, there would be time.

The second letter was from my old and valued farmer friend Will Tinsley. At the beginning of the war some of the best birds from my lighthouse collection had been taken over to his farm to live happily in the orchard and about the farmyard. Among these had been a pair of lesser white-fronted geese, perhaps the most beautiful of all the world's grey geese which I had first met in Hungary and later in their thousands on the Caspian shore.

At that time the lesser whitefront was the rarest British bird; it had only been recorded once, and on any list you cannot have a rarer bird than that. It shared that distinction with some twenty other species which had only been recorded once in Britain. This single record was in 1886 when an immature lesser whitefront was shot in Northumberland by Alfred Crawhal Chapman, brother of Abel Chapman, the famous wildfowler and author. The lesser white-fronted goose breeds as far west as Scandinavia, and from there eastwards across sub-Arctic Europe and Asia almost to the Pacific. Those lesser whitefronts which breed in Lapland fly south-eastwards on their migrations, through Hungary to Macedonia and the Mediterranean.

I had brought some slightly wounded Kis Lilliks (the first word, pronounced Kish, meaning small) back with me from the plains of the Hortobágy to my lighthouse home. When the war came and my collection of live waterfowl was disbanded, a pair of Kis Lilliks was taken over to join Will Tinsley's collection of geese which lived in the orchard beside his house. Now in his letter, Will reported an extraordinary occurrence; he said that in 1943 a wild lesser white-fronted goose had come down one day out of the sky and landed beside his tame pair and had stayed there for several days. There was little chance that the bird could have escaped from any other collection, but as a truly wild bird it was only the second which had been so far recorded in Britain. I had no doubt whatever about the identification. I remember saying to myself, 'If Will says it was a lesser, then a lesser it most certainly was.' From this I fell to wondering how many people there were in this country who would know the difference between a lesser whitefront and an ordinary whitefront.

It is, to be sure, a little smaller but not much; its bill is a good deal smaller and rather pinker, but the only definite distinguishing character is the golden yellow eyelid which encircles the eye of the lesser whitefront. How often, in the field, can a wild goose's eyelids be critically examined? Supposing, I thought, these lesser whitefronts came regularly to the British Isles, who would recognise them? Of course, Will Tinsley would, but I could not think of very many others. And if they came, where should one look for them? It seemed to me that they would be most likely to mix accidentally with those species of geese which migrated to Britain from breeding grounds farther to the east. Two species of grey geese do this – the bean goose and the common or Russian whitefront, and the largest flock of Russian whitefronts in Britain in winter was to be found on the Severn Estuary. Here, then, was the chance of putting my theory to the test. If it was correct I might expect to find a stray lesser whitefront among the larger geese if I could only get close enough to see their eyelids.

A few weeks later I was staying in Stafford and suggested to my friends John Winter and Clive Wilson that we might take up Howard Davis's invitation; on the following day, after a telephone arrangement, we met him at Slimbridge. We walked from the bridge over the canal and down to the end of the lane, after which he led us out towards a war-time pillbox commanding a view of the saltings upon which the geese were feeding. Bent double, we crept across the field, behind the low sea-wall and into the dank concrete box. From the embrasures we had a most wonderful view of a great flock of two thousand wild geese. Among them we saw, quite near by, a young bean goose, then a barnacle, and a brent, and later a greylag. There were also a few pinkfeet but the majority were, as they should have been, Russian-bred whitefronts. That evening we went back to stay with Howard Davis at his farm near Bristol. As he drove me back in his car I outlined my wild idea about the lesser whitefronts and was rash enough to suggest that it was for this very purpose that we had come down to the Severn. On the following day we were back in the pillbox again overlooking the green Dumbles and the grey carpet of wild geese. Again the young bean goose was close in front of us.

We had been in the pillbox, I suppose, for a little over half an hour when Howard Davis said quietly, 'There's a bird here which interests me. Would you have a look at it?' In a few moments he had directed me to the goose in question among the tight mass of geese in front of us, and the instant my binoculars lit upon it I realised that it was a lesser whitefront. My spine tingled delightfully as it does in the slow movement of Sibelius's Violin Concerto. Here almost too easily was a vindication of my far-fetched theory. It was, no doubt, a small recondite discovery, a minor ornithological technicality, yet for me it was a moment of unforgettable exultation – a major triumph, an epoch-making occurrence, a turning point; or is it only in looking back on it that I have invested it with so much significance because, in the event, it changed the course of my life?

From the pillbox we watched the little lesser whitefront for half an hour or more, satisfying ourselves that the eyelids were in fact golden yellow, that the bill was small and extra-pink, and that the white forehead patch rose high on to the crown of the head. The bird had that smooth, dark, perfect look, almost as if there was a bloom on the feathers, which is so characteristic of the lesser whitefront.

Later in the afternoon we moved farther down the sea-wall to get a better view of a part of the flock which was beyond the fence that crosses the Dumbles at the halfway mark. Here among two or three hundred more whitefronts was a second lesser whitefront. To make certain, we went back and found our original bird still almost in the position in which we had left it. Here then undoubtedly were two lesser whitefronts; if Will Tinsley's war-time bird was to be accepted, as I felt sure it should be, these were the third and fourth specimens of their kind ever to be authentically recorded in Britain. It is not often that one sees so rare a British bird and it may be imagined with what excitement we telephoned that evening to a number of ornithological pundits. The meticulous and ever-sceptical Bernard Tucker after much cross-questioning professed himself convinced, but I was privately glad that he saw them for himself a week later. In the following year we saw three lesser whitefronts and in most of the succeeding years in this Severn flock there have been at least one, sometimes as many as six of them, appearing as strays among the Russian white-fronted geese.

On that sunny day in December 1945 the third and fourth lesser whitefronts had brought the total number of kinds of wild geese we had seen together on that marsh to seven, and as we walked back from the pillbox I came to the inescapable conclusion that this was the place in which anyone who loved wild geese must live. Here were two empty cottages which might become the headquarters of the research organisation which had been taking shape in my mind over the war years, the headquarters of a new collection of waterfowl, of the scientific and educational effort which I believed was so badly needed for the conservation of wildfowl. As we squelched up the track, past the hundred-year-old duck decoy, into the deep-rutted yard and back along the muddy lane towards the canal, I looked at my surroundings with a new eye, an eye to the future, for this was the beginning of the Wildfowl Trust.

From *The Eye of the Wind*

20

Bewick's Swans at Slimbridge

The study of individual birds in the wild has been made possible by capture and marking techniques followed by further catching or by close observation, as well as the normal flow of information through recoveries. Observation of breeding birds has been the easiest because of their strong attachment to a limited area. Some work has been done outside the breeding season using conspicuous marks in the form of harness or collar attachments.

Individual variation in birds of one species has been recognised for many years, but it has been exploited for research purposes only in a limited way – for instance, in observing family behaviour in the white-fronted goose, and in the recording of occurrences of the lesser white-fronted goose, at Slimbridge, both concerned mainly with the variation in the black belly bars, and in differences in the shape and extent of the white frontal shield.

The exceptionally favourable circumstances for observation from my studio window of birds on the studio pond in the Rushy Pen (now renamed 'Swan Lake') at Slimbridge have enabled very close observations to be made on wild Bewick's swans (*Cygnus columbianus bewickii*) and considerable variation in their yellow and black bill markings were immediately apparent. The intricate patterns over the culmen and behind the nares form the most ready means of distinguishing individuals, but this can be combined with the underbill colouring and pattern, with shape, size and stance, with eye and eyelid colour, and with other characteristics such as missing feathers, plumage stains, or behaviour. The last are only useful as additional confirmation and of limited value over the long term.

It has been possible to devise a formula by which any Bewick's swan can be described using the variable features and listing them in a conventional order.

Whilst this formula can be used by anyone to describe in letters and figures the facial pattern of a swan that can then be converted into a recognisable drawing, the day-to-day identification of individuals of the Slimbridge flock is nothing like so laborious as the use of a formula might imply. A detailed drawing is used to ensure the correct recording of new arrivals, but once a bird is being seen daily its most obvious features soon assume a prominence in the observer's memory that enables almost instant recognition of any bird. Additional items are, of course, used in confirmation in the daily recording, one of the most useful here being the presence of a mate or family.

It has proved possible for someone coming new to the swans to master the characteristics of the birds quite quickly, assisted by the drawings.

The face patterns of cygnets present additional problems, particularly early in the winter. The black markings appear progressively over the reddish pink areas of the bill and the limits of the yellow are ill-defined at first. The drawings are made of them as late in the season as possible to get the best picture of what they may look like the next year. In their second winter it is usually obvious from a bird's behaviour whether it has been to the pond before (new birds start up-ending for food in deep water and away from the feeding place). There is a tendency for siblings to associate, and also to consort with their parents, even if new cygnets of the year are also with them. All these points help to confirm identification if the bill pattern is not immediately recognisable. There is no doubt that this is a possible difficulty in future. Another difficulty is that the patterns of the adults are subject to minor changes from year to year. Some change more than others, and the changes may involve more yellow or more black.

In only three winters' experience of these changes no system of prediction has emerged but after a further period it may be possible with greater knowledge of the physiology involved to discover certain rules governing the changes.

Bewick's swans were not very frequently recorded on the Severn Estuary in the early years of the Wildfowl Trust, but by 1956 the occurrence of the species had become regular and up to 16 birds were visiting the Big Pen in the late winter. The numbers increased slowly until by the winter of 1963–64 more than 20 were present for a considerable period. In February 1964, by dint of moving 3 pinioned Bewick's swans and 4 pinioned whistling swans (which the wild Bewick's do not apparently differentiate from their own race) into the Rushy Pen, 24 wild birds were persuaded to come regularly to the pond in front of my studio window where they were liberally fed. In the following winter of 1964–65 there were 55 wild Bewick's at the peak time, and 68 were recorded during the winter. During the winter of 1965–66 the pond for a considerable period held over 120 swans with 125 as a peak; and 147 different swans were recorded in the season.

Detailed drawings of the facial patterns of 188 different swans have now been made, and each bird has been named for easier reference. The names are used as aide-memoires in identifying the birds and may refer to a particularly well-marked facial characteristic (Two-spot, Shieldy, Freckles), to unusual colouring (Lemon, Amber, Pink) or to a behavioural feature (Caesar, a very aggressive male).

Given the possibility of individual identification it follows that after the second complete winter of the study we must already have learnt something. The most important fact so far is the confirmation in yet another species of Anatidae of the strong traditions inherent in the birds in their choice of winter quarters. This is borne out by the return year after year of the same birds, bringing their young of the year, thus leading to a part of the increase taking place, and guaranteeing a continuing increase in years to come. However, the arrival here of adult birds that have not previously wintered (and these form a considerable proportion of the increasing flock) does argue that this tradition can break down under certain circumstances. What these circumstances are we can only guess at, though we would perhaps be justified in thinking that the unlimited food (wheat) – and to a lesser extent the security – at Slimbridge must be the principal attractions. We cannot be sure how these attractions are discovered by new swans, though we suppose that some attach themselves to swans who know the place at earlier staging posts during migration. Others meet up at nearby places when our birds wander and are subsequently brought in by them. A third method may be the actual sighting of the flock by passing migrating birds.

The return of birds in succeeding years has been most encouraging. Even before the start of facial recognition we had evidence of this from ringing. An adult caught in the pens on 2 April 1961, was subsequently recaught on 21 November 1961, and again on 10 February 1963. Of the 24 birds identified in winter 1963–64, 16 returned during the next season 1964–65. These were five established pairs, one of which brought two cygnets, and another which brought three of the previous year's four cygnets, having apparently failed to breed in 1964. The single young of the only other successful breeders of 1963 did not reappear. Of the three unattached birds of 1963–64, which returned, two brought mates with them.

The first birds to arrive in 1964–65 were a pair (Pink and Rebecca) that had been here the previous year, and they brought two cygnets with them. This was on 4 December. The next arrivals were not for a fortnight, and then all the next eight (Maria, Pop, Mom, Ranger, Sis, Big Bro, Owl, Pussycat) were from 1963–64. In early January some new birds started trickling in, not all staying very long. Some longer-lasting influxes took place in the middle of January, and in early February, with up to 55 swans regularly using the pond. A mass departure took place on 15 and 16 March 1965, a single family (Pink and Rebecca) hanging on for another week. Altogether 52 swans were identified in addition to the 16 from the previous year, so that the total number that came to the pond was 68, 13 above the maximum reached on any one day.

On 21 October 1965, the same pair that arrived first and left last the year before (Pink and Rebecca) arrived on the pond with three cygnets, together with a single new adult. They were six weeks earlier than in the previous autumn. In the next three days 14 more swans came: two pairs (Kon and Tiki with three cygnets, and Pepper and Amber with two cygnets) that had been last year, a single old adult – originally paired but a widow during the previous winter (Maria) and four second-winter birds, all of which had been before as cygnets. Two of these were Reuben and Rachel, the young of Pink and Rebecca in the previous year. They immediately joined up with their parents and the new cygnets making a flock of seven. By 1 December, 86 birds had arrived and stayed, of which 27 were cygnets (one of these was killed in November flying into a tree). Of the 59 adults or two-year-olds, 29 had been in 1964–65, and 11 of these in 1963–64 as well. During the rest of the winter a further 61 swans, of which 15 were cygnets, came for shorter or longer periods, but only four of these had been in 1964–65, and none from 1963–64. The maximum on any one day was 125 and the total for the season 147. Thus there is a very strong tendency for the birds, having once learnt of the place, to come here fairly early in the winter.

The duration of stay of the swans varied in the two seasons, mainly because of the greatly differing dates of arrival. In 1964–65 most of the birds stayed in the area once they had arrived, occasionally missing a day in visiting the pond for food.

The main exception to this pattern was the first family to arrive (the Pinks) which came on 4 December, departed on 19 January and did not return until 9 February. They then stayed without a break until a week after all the other birds had departed. Eight single birds that arrived in late December or early January stayed for short periods only, from one to 27 days, one other stayed for 17 days, departed for three weeks, and came back for two days more before departing for good.

The longest continuous stay was 87 days (by Maria) with several more families or individuals between 60 and 80 days. The mean stay of all 68 birds was 48 days, and of those which stayed until the general departure in mid-March, 55 days.

In 1965–66, there were two important differences from the previous winter. First, the arrival of the birds started six weeks earlier, 86 birds having arrived by early December, and secondly, the very wet period in December with widespread floods in the Severn Valley and elsewhere caused many birds to depart for up to four weeks; though, with the exception of two birds (Romeo and McJuliet) which never reappeared, they had all returned by the second week in January when ice covered many of the flood waters. There were other shorter or longer gaps in the attendance on the pond of some other birds, but none more than a few days in duration, and not necessarily indicating that the birds had wandered far. The goose shoots in January had temporary effects of this nature.

The longest continuous stay was by the Pink family, which had again been the first arrivals and this year had scarcely an interruption in their stay, even during the floods, though they were missing for some odd days at this time. They stayed in all for 154 days. There were many more stays of over 100 days. The mean of all 147 swans was 79 days, and for those which stayed until departure time in March the mean was 89 days. Thus the swan usage of the pond was very considerably higher than in the previous winter, roughly 11,700 days compared with 3,250 days, or an increase by a factor of 3½.

The maximum numbers of swans regularly using the pond over a period also went up enormously, from 50–55 maintained for the period 11 February to 15 March 1965 (33 days), to 110–125 almost continuously from 19 January to 10 March 1966 (51 days). This kind of usage of a small pond raises the question of when the birds will begin to feel overcrowded. The other more frequent limit of numbers, namely food supply, will not arise in this case as the amounts of wheat fed twice a day can be increased in proportion to the number of birds. For 120 swans 2½ cwt of wheat were fed daily, with a much smaller amount of biscuit meal. There is some observational evidence that the more birds there are on the pond, the greater the frequency of aggressive encounters between families or individuals, but so far there have been few records of birds being prevented from reaching the ample food supply because of the presence of either too many or too aggressive birds. It may be necessary in the future to enlarge the area over which the food is scattered, but the water area now seems adequate, both for landing and taking off (for which it has been enlarged already) and for the normal bathing and resting activities. It seems possible that 'Swan Lake', as it is now called, could accommodate up to 500 Bewick's swans.

We have so far ringed 23 Bewick's swans, catching all of them in the pens, usually when they have got into a confined space where take-off is difficult. Two recoveries have been made away from Slimbridge. The first was an adult male caught on 10 February 1963, and kept in the pens until the following September. It was found dead on 19 May 1964, in the Nenetsk National Okrug, USSR, within the known breeding range of the species and about 2,300 miles from Slimbridge. The second recovery was of Elmer, who arrived on 11 January 1965, and was ringed the next day when he hit a chimney and fell into the garden. Although injured at the time, he made a complete recovery and departed on 16 March with his mate (Petula). He was found dead at Frodsham, Cheshire, about 25 November

1965. His mate has not returned. It is hoped in future to use rings with large numerals so that they can be read with binoculars and positively confirm identification of birds on the pond.

The behaviour of the birds and the relationships between them are being closely studied. The pattern of aggression between families and individuals is noted each time it is observed, and a nominal order of dominance is being drawn up. Already there are nearly 200 records of aggressive encounters and a peck order can be worked out. As in the geese, the larger families tend to dominate the smaller, but not so rigidly, and a really aggressive male (such as Caesar) with only two cygnets can defeat parents of larger broods. Pair-formation behaviour has been noted many times, and in 1965–66 seven pairs were apparently formed during the winter. The two-year-old birds spend quite a lot of time in courtship display, often to different birds within a few days. There are some indications that birds pair with others having a similar bill pattern to their own, but this must be the subject of further work, possibly using the formula method of description. The inheritance of the various features of the bill pattern is also under study, and so is the degree of change from year to year in each bird's pattern.

We believe that the discovery of this method of individual recognition may in the course of a few more years lead us to a number of new discoveries about the biology of the Bewick's swan.

From 17th Annual Report of the Wildfowl Trust

21

How Much Does It Matter?

All living animals (including man) are the current 'end-products' of twenty million centuries of evolution. Since life began new kinds of animals have been constantly coming into existence by a combination of the processes known as 'adaptive radiation' and 'natural selection', while other kinds, unable to cope with changing conditions, have become extinct. The existence of a tiny proportion of those extinct species is known to us from fossils. During the last two hundred years – geologically speaking, a mere instant of time – the direct or indirect impact of man's progress has more than doubled the rate of extinction of species and the process is still accelerating.

At the present time over a thousand kinds of vertebrate animals, species and races, are in danger of becoming extinct at the hands of man. They are threatened by greedy and improvident short-term exploitation, or by unwitting carelessness and apathy, because they do not seem to have a 'use' and their disappearance is therefore held to be of no consequence. Some are regarded, more often wrongly, as injurious to man's interests and are being wiped out deliberately and systematically. It is fair to say that human ignorance is the basic cause of the threat in nearly every case.

In such a situation there are inevitably a number of moral and ethical considerations. Many people consider we have a clear moral responsibility to keep 'a place in the sun' for the animal species which share the earth with us. They see this as an issue of right and wrong in which it is wrong to eradicate any species. Others feel that the extinction of a species could be justified, but only by some powerful and morally compelling reason such as the avoidance of large-scale human suffering.

In this context it is interesting to note that not one animal which has become extinct within historical times, such as the dodo, the quagga, the great auk, the passenger pigeon, the blue antelope, the Labrador duck, the Carolina parakeet (and about a hundred more) could have delayed the progress of the human race in the smallest degree, had it survived.

Another ethical argument is that we are answerable to future generations for what we do to the natural world around us. Just as we have no right to allow radioactive fall-out to prejudice the health of unborn babies, so we have no right to destroy the natural world with its flora and fauna which is the rightful inheritance of generations to come.

Whether or not these arguments carry weight, it is clearly indefensible to allow something so irrevocable as species extinction to occur as a result of ignorance, apathy, or bad judgement.

There are powerful aesthetic arguments in favour of saving the world's wildlife and wild places. People *enjoy* animals, find them beautiful and exciting to watch, and derive spiritual refreshment from their study.

There are scientific reasons in that much still remains to be learned about animals. Not only may it be considered wrong to remove potential subjects from the field of zoological research, but at the more material levels of human progress it is foolish to wipe out an animal which might later turn out to be of practical use to mankind.

There are also economic arguments. In many parts of the world wildlife is a major tourist attraction and this is capable of wide extension. With advancing education, and cheaper travel, more and more people will go to see the wonders of wildlife in distant lands.

Wildlife can also be 'farmed' as a source of human food. Already schemes have been successfully developed for cropping surplus animals without endangering the breeding stock. In many areas of marginal agricultural value wild animals can produce a far greater yield of protein per acre than domestic animals; they are less susceptible to disease and much less likely to damage the range by over-grazing. Whether as a tourist attraction or as a renewable source of protein, wildlife very often offers the most economical as well as the most enlightened kind of land use.

It should not be forgotten that man's own survival depends not only on his capacity to avoid blowing himself up, but also on his capacity to stop making deserts and dust-bowls, to preserve and replant forests, to husband the earth's resources of animals and plants. Man must learn to apply the science of ecology to himself. His future is bound up with his right assessment of the use of land. In this context alone it seems clear that he has responsibilities of trusteeship for the natural world on which his activities have such sweeping impact. He may not yet be able to limit his own increasing over-population, but nuclear fall-out, oil pollution of the sea, freshwater pollution, toxic chemical sprays, and greedy over-exploitation of natural resources are factors which he should be able to control, and surely could if he thought it sufficiently important.

It may be asked whether at this time there is room to feed the ever-increasing human population and at the same time provide adequate space for wildlife. The answer is at present an unequivocal 'yes'. Science *can* keep pace, if politics will let it. With foreseeable advances in the production and distribution of food, in the productivity of land and sea, there is no sufficient reason for the indiscriminate destruction of wilderness and wildlife habitat which still goes forward all over the world without proper technical advice and forethought.

As a subject for charitable appeal, how does the saving of wildlife and wild places compare with the more usual charities, most of which are devoted to the relief of human suffering? It is generally accepted that we have a first duty to put right present human misery, but there is also future human misery to consider, and a duty to take the long view of the use of land, so as to avert future disaster.

At least as important is the recognition of responsibilities at the other end of the scale – in educating, in enriching human lives. Although we live in a grossly material world, it is still true that man cannot live by bread alone.

'Reverence for life' is not widely practised as a doctrine, though many great thinkers have accepted it. But the ethics of taking animal life should not be confused with the ethics of exterminating species. Sportsmen and non-sportsmen do not agree on the first issue but usually agree absolutely on the second. There is in any event a strong case, when wildlife conservation is a matter of such extreme urgency, for those who think killing (and especially killing for sport) is wrong to make common cause on *this* issue with those who do not, but who are still a majority of mankind. It is doubtless good sense for *all* who understand the urgency of the present crisis to combine together in a common crusade.

Concern for the continued existence of a rich variety of wildlife and wild places is as surely a manifestation of human enlightenment as are art, discovery, exploration, education and tolerance.

From *The Launching of a New Ark*

22

A Voice Crying in the Wilderness?

With thoughtlessness, cruelty, and greed man destroys the natural world around him, too often quite unnecessarily. But the technical progress which enables him to do this so sweepingly is matched by an awakening of conscience that perhaps this same man, as the first predatory animal able to reason things out, has some responsibilities of trusteeship for the earth he lives on.

Causing the extinction of the dodo, the great auk, the passenger pigeon, the quagga, and hundreds more (at the rate of at least one animal form per year during this century), and threatening the extinction of about a thousand other kinds of animals (including the blue whale, the largest animal so far as scientists know that has ever lived), may or may not be morally wrong. But the conservation of nature is most important because of what nature does for man.

I believe something goes wrong with man when he cuts himself off from the natural world. I think he knows it, and this is why he keeps gardens and window-boxes and house-plants, and dogs and cats and budgerigars. Man does not live by bread alone. I believe he should take just as great pains to look after the natural treasures which inspire him as he does to preserve his man-made treasures in art galleries and museums. This is a responsibility we have to future generations, just as we are responsible to them for the safe-guarding of the Parthenon or the *Mona Lisa*.

It has been argued that if the human population of the world continues to increase at its present rate, there will soon be no room for either wildlife or wild places, so why waste time, effort, and money trying to conserve them now? But I believe that sooner or later man will learn to limit his own population. Then he will become much more widely concerned with optimum rather than maximum, quality rather than quantity, and will rediscover the need within himself for contact with wilderness and wild nature.

No one can tell when this will happen. I am concerned that when it does, breeding stocks of wild animals and plants should still exist, preserved perhaps mainly in nature reserves and national parks, even in zoos and botanical gardens, from which to repopulate the natural environment man will then wish to recreate and rehabilitate.

These are my reasons for believing passionately in the conservation of nature.

All this calls for action of three kinds: more research in ecology, the setting aside of more land as effectively inviolate strongholds, and above all education. By calling attention to the plight of the world's wildlife, and by encouraging people to enrich their lives by the enjoyment of nature, it may be possible to accelerate both the change in outlook and the necessary action.

It has been estimated that conservation all over the world needs each year about £2 million. This is no astronomical figure. It is half the price of a jet-bomber, less than one-twelfth the price of a new ocean liner, or the price of, say, three or four world-famous paintings.

Much money is needed for relieving human suffering, but some is also needed for human fulfilment and inspiration. Conservation, like education and art, claims some proportion of the money we give to help others, including the as yet unborn.

Even if I am wrong about the long-term prospects – if man were to fail to solve his own over-population problem, and reaches the stage when there will be standing room only on this earth (one square yard per person calculated at 530 years hence), even then the conservation effort will have been worth while. It will have retained, at least for a time, some of the natural wonders. Measured in man-hours of enjoyment and inspiration this alone would be worth the effort.

But I do not believe this will happen. Over-population will be solved, and I believe mankind will recognise the importance of wild nature to his own well-being long before he has destroyed it all. The community chest which seeks to bring massive support to conservation all over the world, and to make people aware of the problems sooner rather than later, is the World Wildlife Fund.

23

The Best the Arctic Can Offer

Tuesday 19 July. In the early morning there were two lesser Canadas, one lesser snow, and six Ross's geese, all flightless, in the pool below us. But it was useless to attempt to catch them. There were also about fifteen pintails.

As we were having breakfast, we suddenly heard a Ross's goose call upstream, and I saw a party of about fourteen of them coming over the hill, doing a portage, as it were, round the rapids. They came on towards us, and we waited. We thought for sure that, like yesterday's party and this morning's six, they were flightless birds. At last, grabbing the hand-nets, we rushed to cut them off from the river. There was a steep gravel slide, and they started to climb this and then to fly off it. Too late, we realised they all had young with them, which were rolling and tumbling down the gravel. At the bottom of the slope was a little pool. The young collected there. They had evidently taken no harm from rolling down the steep slope. We noticed that some were silver and some gold. There seemed to be rather more silver ones, but we did not get a chance to make a count of them.

One brave pair of adults flew in and settled on the pool. Presently they took off again, and went over the hill out of sight. The young were scattered all over among the boulders, and, circling above, were a glaucous gull and a herring gull. We were extremely worried. After a while things began to straighten themselves out. A couple of pairs flew over on to the gravel slide and collected two goslings. A group of four goslings came down past us to the river, and two more joined an odd one in the pool below the rapids. These seven were unattended, but four of them stayed near us and were therefore safe from the gulls. I got out the gun to shoot the gulls, but it did not seem to be necessary, and indeed once I had the gun they did not come close enough. Two pairs of Ross's kept flying over the hill and then turning back upstream again. At last, one pair flew past us downstream, and then suddenly saw the flotilla of four goslings below – landed and took charge of them. Then the second pair flew over, and a few minutes later I saw the three with them. So, all was now well.

As if to show that we had not so seriously disturbed them, another pair began walking over the portage with their brood of four. They came by a different route, but they came close past the camp. I got out the telephoto movie, and made some pictures of them as they passed along a snow-bank, dropped off into the river at its edge, and swam away downstream. At a hundred feet the family filled the screen.

Now, I am up-to-date with this journal, except to record that a ground squirrel appeared on the bank just now and let out a series of shrill whistles, which we have not heard before.

Before we set off on the portage over the next set of rapids, I must draw the little Ross's goslings.

Having drawn one gosling, we set off. The first part of the journey was fairly wretched. We portaged round about a mile of rapids. With only two of us, this meant three journeys because of all our gear. On the second journey, I carried the canoe. This meant very nearly six miles of walking, half of it heavily laden. The weather was perfect, neither too hot nor too cold. The mosquitoes were only intermittently troublesome.

Having completed the portage, we had lunch. After lunch, our day began, and from then until midnight we had the most wonderful time since we landed on the Four Lakes five and a half weeks ago.

The rapids continued and progress was very slow, pushing the canoe among the rocks, but anything was better than carrying the gear. Some stretches we covered faster, with a bow and stern line and a technique which we have now perfected of easing or tweaking the ropes so as to steer her clear of the rocks in the stream. It was still tough going for Harold, as his foot was giving trouble, but he had bound it tight and could 'get by', so he said.

Moving upstream ahead of us was a group of about nineteen pairs of geese with young, mainly Ross's, but also one pair of lesser snows, with three, and one pair of lesser Canadas, probably the pair I had seen the day before. Many of the Ross's were no doubt some of the party we had surprised while making the portage round the rapids lower down. At one corner we stopped and made brood counts of the young. These came out, as follows, on the thirteeen nearest broods: 3 yellow, 2 grey, 3 grey, 1 grey, 3 yellow, 2 yellow and 1 grey, 1 grey, 4 grey (very pale), 2 grey (pale), 4 grey, 2 grey, 3 grey, 3 grey (very pale). In 34 goslings – 8 yellow; only one mixed brood, and average brood size 2.6.

Then we climbed up the bank of the river, and found a lake, which was – as it were – the top of the ridge. Harold christened it 'Skyline Lake'. It had a pair of lesser snows, with a brood (3) and three pairs of Ross's on it, also with broods (3, 3 and ?).

Flying over the lake was a raven, only the second we have seen here. There were also six long-tailed ducks, and a glaucous gull.

On a nearby mound was a ground squirrel, with a family of about six. She was carrying them down the hole in her mouth. I came to within twenty-five feet with the movie camera and got a lovely shot of the mother with two of the young. How I hope we shall find we have been giving the right exposures and judging the distances correctly for these big lenses. I reckon that a third will be exposed wrong, and a third will be out of focus. If the other third is all right, there should be material for a good film. So far, I have taken more than 3,000 feet (16 mm).

Looking across the river, we could see another lake with white geese on it. The river curved round towards it, so we worked the canoe up until we had come to the nearest point, about 300 yards away from it. Then we took the cameras, and stalked up to the edge and looked over. It was a larger lake than Skyline Lake, and was about three-quarters of a mile long. Just in front was a group of geese, evidently non-breeders but not yet moulted. There were eleven Ross's, four lesser snows, and an adult blue goose. On the far side of the lake was a galaxy of fowl: fifty-five long-tailed ducks, nearly all males, all in a bunch, and possibly flightless. In addition, there were twenty others scattered about the lake diving for food. There were seven female king eiders, and they were strangely curious, swimming over to look at us. On the far shore were six pintails, including a drake still in full plumage (all the others were in full eclipse). Swimming down the far shore were two families of lesser snows, one with three goslings, and one with two; and up to the left, where the blue water ran away to our skyline, were two families of Ross's. One had three yellow and one silver-grey gosling, and the other had two yellow and one grey.

Harold named this 'Highland Lake'. I cannot describe its beauty in words, though I shall do paintings of it. The water was unbelievably blue; the hills were the softest lavender-grey to the south of it. To the north it seemed to be at the very lip of the high ground to which our flights of rapids had brought us. Like a brimming saucer, its northern shore reached our horizon. Beyond, the low rolling tundra dropped away to the plain of the Perry. Harold said he felt as though 'he were at the top of the world – almost in the sky!' He waxed quite lyrical about it, and I with him. This was what we came for. This was worth all those first cold, dreary weeks of snowstorms, worth the anxieties of the flight, worth all the frustrations – a million times worth the rapids and portages of the trip, so far.

We felt fairly sure that these geese had been hatched on the islands in Lake Arlone. Some were obviously almost a week old. Evidently, they spread out by swimming down the river to these lakes and green banks, and farther down to the lush prairies where we had met the little party portaging round the rapids.

We walked out to a peninsula on the lake, hoping to find perhaps a hatched goose nest, or at least a long-tailed duck sitting, but there was nothing. On the way back, we stopped and watched a semi-palmated sandpiper go back on to her nest. She still had four eggs, although I am sure that some I have seen lately have had young. These tiny little birds are probably the second commonest waders up here (the pectoral sandpiper being the commonest). As we got back to the river the seven female king eiders flew past us and settled just ahead. They kept us company for the rest of the day.

Round the next corner we found a brood of tule geese. Since these were apparently the first downy goslings of the tule goose ever to be seen, or, at any rate, recorded, it was evident that they would have to be sacrificed to Science. I will not dwell on the horrible business of trying to secure the parents, of catching the goslings, of releasing one in order to try to decoy the parents, and, finally, losing the one which had been released, so that four tule goslings were all that the scientists finally collected. By the time it was over it was evening, and we stopped for tea which consisted of biscuits and jam, and some cocoa-bricks eaten like chocolate and washed down with milky river water. We reckoned that the river could not climb much farther, and indeed soon we came to a long flat stretch along which we could paddle. Here was the best sight of all. The families of geese which we had been driving before us had joined a whole lot more. There was a great collection of geese with their broods. We followed some of them in the canoe. One brood of three had a runtling which lagged behind. For its sake, we by-passed that family and went on to get photographs of the little flotillas of goslings, each with the two parents shining white in the late evening sun. It was a scene of miraculous beauty.

We landed on a point. Beyond it were yet more geese. We lay and watched with glasses. The geese were moving upstream, but now that we were more or less out of sight they went at a leisurely pace. There were sixty-four adult Ross's geese – thirty-two pairs with their young; one pair of lesser snows; two pairs of lesser Canadas, each with four goslings, and a pair of tule geese, also with four goslings. With them were our seven female king eiders, who had been joined by two more, and also there were two drake king eiders, still in full breeding plumage. A few long-tailed ducks, a glaucous and two herring gulls made up the picture.

We camped on this point, with a calm stretch of the lovely River Kennet above and below us, and the tinkle of a minor rapid opposite us. There was a glorious prolonged sunset as we ate our supper at about 11.30. We went to bed – exhausted but content. We were, we felt, achieving some of the main objects of the expedition – seeing new things and making new discoveries which were of real importance and significance. No ornithologist had ever before seen Ross's geese at this juncture of their lives.

Wednesday 20 July. Another glorious, sunny day with a light northerly breeze. Part of the success and enjoyment of this trip up the Kennet has been the perfection of the weather. I have been writing up this journal lying in the sun, while Harold skins the last of the Ross's goslings and a ground squirrel which he shot yesterday morning. He still has a herring gull, which he shot last night, and the four tule goslings to do. But he may leave them till later, as it is almost noon and we are anxious to push on in the happy – though optimistic – belief that Lake Arlone is only just around the corner. Actually, I think, it may well be another six or seven miles. But after all, that is the delight of exploring totally unmapped territory.

One wonderful day succeeds another. It is 11.30 in the evening of the most exciting day of all. We have been among the geese all day, learning new things about them, as we advanced into new and exciting territory. We set off along a level stretch of the river, but very soon we were back on a group of rapids and working up them in the usual laborious way. Ahead of us once more was a group of geese, with broods. There were about twenty-five pairs, and in a while we noticed that they were collecting in a pool at the head of which was a rapid, up which they could not go. The river took a very sharp bend at this point, and the far shore consisted of steep rock faces and snowdrifts. In short, the geese were cornered.

First of all, we made a count of the broods, with their colours, which came out as follows:

Ross's geese: 2 grey, 1 yellow; 2 grey, 2 green, 1 yellow; 2 grey, 1 yellowish; 3 grey; 2 silvery grey; 3 greeny; 1 green, 1 grey; 3 grey; 4 grey; 2 grey; 1 white-yellow; 2 grey, 2 white; 3 grey; 1 yellow, 2 grey; 3 green-yellow; 1 dark grey; 2 silvery white; 2 dark grey; 2 grey, 1 yellow; 18 broods, average size 2.8.

In addition to the Ross's there were two pairs of lesser snows, with broods of three and one respectively; and a most interesting pair of which the female was a lesser snow, the male a blue goose with white flanks and belly. The two young were typical lesser snow downies – greeny-yellow.

Having counted them, we got into the canoe and paddled out to them, photographing them as we went. If I judged the distances right, we should have some splendid material, but it was difficult as, in the swirling current, the distances were changing all the time. The birds stayed with their young and allowed us to paddle up to within about seven or eight yards.

After completing our pictures, we went up the rapids and found the beginning of a long lake stretching away in front of us. At first we took this for an arm of Lake Arlone, but later we realised it must be another lake, and we called it Jane Lake. It is at least four miles long, and for the most part no more than a quarter to half a mile wide. All along the eastern shore there were Ross's geese with their broods, and occasionally there were lesser Canadas and lesser snows. In one or two places were groups of flightless geese, without broods. On land these could be spotted at once, as they ran much faster, not being encumbered with goslings.

At one point we stopped to 'collect' a pair of lesser Canadas and their brood of two goslings. Harold did the distasteful job, while I picked up one lesser snow gosling from a brood of five. The lesser snow parents were brave and came back to the gosling when I overtook them. I ran back to the canoe for my telephoto lens, but the shot which I finally got was not very satisfactory.

Ahead of the lesser snows was a pair of Ross's, but when I overtook the lesser goslings I found a Ross's gosling running just behind them. There is no doubt that we must do as little of this pursuing as possible, for it *does* break up and exhaust the broods and might mean the loss of goslings.

Paddling south-westward along the lake, we counted each Ross's brood as we came level with it, and, as the light permitted, we made an estimate of the colour-phase of the young ones. We were so engrossed with this that we passed by the mouth of a tiny stream which we should have taken, and proceeded along the narrow, and now shallow, upper arm of Jane Lake. Ahead we saw small groups of moulting lesser Canadas, and a large group of at least fifty moulting snow geese. We paddled on for some time before the snow geese saw us. When they did, they suddenly ran together into a tight bunch and went into the water. We overhauled them fairly rapidly in the canoe. When they were about fifty yards ahead, they began to flap along the surface. There seemed to be geese in all stages of moult. One or two could fly and took off; some had old feathers, but not enough left to get into the air; others had half-grown new primaries; the majority had no primaries showing at all. As we passed through the patch of water where they had done their first flap, the surface was covered with quills which had just fallen out. The geese swam in a solid phalanx, no water being visible between them. We were now close behind them, about five yards away. They were mostly Ross's, a few lesser snows, three blues, and one lesser Canada – which went a little faster than the rest and became the leader of the band.

One Ross's got left behind, and we by-passed him and went on for the rest. I took what should be an excellent movie shot of them just in front of us. We drove them ashore – in the wrong place, on a point with water beyond. Running, they went much faster than on the water and got well ahead of us. Also, the lake had become a river again and the stream was against us. About a dozen broke away and went out over the tundra; we decided to try for them, although they already had a hundred yards' start. We should probably have done better by going on after the party which had taken to the water again beyond the point. As it was, we ran all out over the tundra for about 400 yards, Harold oblivious of his bad foot. The birds began to crouch when we overtook them, and eventually we picked up four Ross's and one lesser snow. One Ross's was a female, apparently in breeding condition, and all the rest were males – the lesser snow being a yearling.

After the capture, we sat exhausted on the tundra and then slowly made our way back to the canoe. Here Harold ringed them, and I filmed the operation and the final release. We think that these are probably the first Ross's geese to be banded, with the exception of some wing-tipped ones in California.

We decided to cross the lake and climb a hill, so as to see where we were, for we were convinced that we could not be far from Lake Arlone.

From the top of the hill we could see a lake, but at first sight it did not appear to be Lake Arlone. In the opposite direction, to the south, the Kennet continued out of the head of Jane Lake and ascended a fresh flight of rapids. Beside the rapids were six caribou. They were two miles away.

We walked farther on to the northward to another peak, seeing new flowers in bloom, including this time the real rhododendron; and from this new vantage point we could make out the island in the lower half of Lake Arlone. We were there! And Harold and I solemnly shook hands upon it.

We paddled the canoe back down Lake Nicola and found the tiny river leading out of the other lake, which we had missed in our zeal to count all the Ross's broods. We made camp at the mouth of this river, for we thought it might disturb the birds were we to camp on Lake Arlone itself. The little stream, which connects the two, is not more than a couple of hundred yards long.

After supper, we walked to the top of the hill behind our camp. There was a glorious sunset coming across the still waters of Lake Arlone. Little flotillas of Ross's geese, with their broods, left wakes across the water. On the lower island – the only one we could see – at least four females were on their nests still. We could see three of them and the gander of the fourth. So we may get a picture of a sitting Ross, which is what we want. Some of the families, which had been at the foot of our hill and had swum out when they saw us, went across to the island where at least one female took her young back to the nest and brooded them there.

Harold stayed up on the hill to get a picture of the lake named after his wife, when the sunset was at its best.

Back at camp, there were some pintails, still able to fly, on Jane Lake, and some flightless lesser Canadas and half a dozen families of Ross's. It is evident that this movement down Jane Lake to the Kennet, and down the Kennet, is a traditional one for the Ross's goose broods from Lake Arlone.

The northern sky was golden green, but to the south, and reflected in the incredibly still water of Jane Lake, the sky was a soft lavender colour. Harold was greatly impressed with its beauty and called me out of the tent to have a last look at it. In the last two days we have touched the top; we have experienced, surely, the best the Arctic can offer, and surely, too, the best that any explorer – even amateurs, as we are – could possibly wish for.

24

Wild Music

Sound has a powerful effect on man's emotions, perhaps more powerful than sight, and whether the sound is of a man's designing or of nature's, it may bring us very near to tears by its beauty. Heard above the rushing of the wind, the cry of wild geese can be overwhelmingly sad.

The nightingale and the blackcap and the curlew are nature's soloists but the geese are her chorus, as rousing, over the high sand, as the 'Sanctus' of Bach's B minor Mass. As they flight at dawn one can imagine that each successive skein brings in the fugue, '*Pleni sunt coeli . . .*'

This very autumn of 1935 I saw and heard the pinkfeet arriving from the north.

We had been sitting all day in the punt. For a while the rain had ceased and there was a light patch in the watery sky, that seemed to be scurrying before the next black squall. Suddenly a goose called. Very high up in the light patch of sky were about thirty geese with set wings, swinging round on the wind. They dropped down in a great sweeping curve, blown like autumn leaves to the far side of the estuary and then working back low to settle on the top of the high sand.

All that day they arrived in skeins of twenty and thirty and fifty, and all the next day too, until there were five thousand geese on the high sand.

They have come south again to exist against incredible odds

in a land of human beings, until the season of midnight sun thaws out their northern breeding-places. The marshes will be filled with their unparalleled music as they flight at dawn and at dusk. When the moon is full they will pass unseen in the steel-grey sky to their feeding grounds, but their cry will echo across the flat fields.

Like a symphony of Beethoven, the call of the geese is everlasting, and those who have once known and loved it can never tire of hearing it.

From *Morning Flight*

25

South America – 1953

In February 1953 I left Slimbridge with my wife on an expedition to South America. The objects of the expedition were, first, to see in their natural surroundings some of the common waterfowl of the region, which have been kept in captivity in Britain for many years; secondly, to see some of the lesser known species, such as steamer ducks (*Tachyeres*) and kelp geese (*Chloephaga h. hybrida*); and finally (and most particularly of all) to see and to study three species about which very little indeed is known. These were the black-headed duck (*Heteronetta atricapilla*), the bronze-winged duck (*Anas specularis*) and the torrent duck (*Merganetta armata*).

All these objects were achieved. We travelled by way of North America and the journey from there to South America was undertaken in the company of Col. and Mrs Goss and two other members of an expedition from the Museum of Natural History of Cleveland, Ohio – one of them its curator, Mr Rendell Rhoades.

We had two days in Santiago and one of them was spent in going down to the coast to see some birds. We were taken by A. W. Johnson, the distinguished ornithologist and one of the authors of *The Birds of Chile*, and accompanied by J. D. Goodall, the illustrator of that excellent book. They have built up the first really good collection of Chilean birds' eggs. There is no question that this thirty-year task is of the greatest scientific value. They have been the first to find the nests of dozens of species. We saw their collection, including several sets of coots' eggs, with black-headed ducks' eggs among them. They have also found the black-head parasitic on night herons and the ibis. In one clutch there were four ducks' eggs among the coots' eggs, while several clutches held more than one duck's egg. Are these laid by the same female? Johnson was convinced that the black-headed duck *never* incubated its own eggs.

They took us first to a pond called El Peral (the Pear Tree) near the coast – indeed, just inside the sand dunes, near Cartagena, where they had found eggs of the black-headed duck. On the pond we found Chilean or brown pintails, Chile teal, and two pairs of cinnamon teal, which looked marvellous in flight. We also saw some of the narrow-billed ruddy ducks. These are just like the North American ruddys, but with more slaty grey-blue bills (perhaps because of the season) and no white cheeks, i.e. somewhat inferior birds in appearance.

We moved to another lake thirty miles away, to the south – a much larger one – where it was hoped we might see some ducks and black-necked swans. This lake was on a farm called El Convento. We reached it along a track which led through sun-scorched fields, full of tall seeding thistles as big as artichokes, down towards the wooded shores. At the end of the track was a eucalyptus belt, and we looked out through it on to the lake. In front of us was a large red-necked grebe – a great grebe – with a young one. There were coots and moorhens, and one or two ruddy ducks up under the sun. Across the lake in the floating weeds were either brown pintails or Chile teal, but to the right up at the narrower end of the lake was a great mass of ducks among the floating weeds. At first glance I could identify rosybills, chiloe wigeon, and red shovelers.

We decided to approach these ducks along the shore, and look down on them from the cover of the eucalyptus belt, and eventually, having shown ourselves inadvertently to the nearest birds, which swam out, we took up our position with the camera tripod for a rest and began to scan the flock with binoculars. For the next hour we had as thrilling a view of new and exciting birds as I can remember for many years. There were at least two thousand ducks below us, scattered, feeding and resting among the floating weeds. It was such a view as we have so often had along the shore opposite St Serf's Island in Loch Levin in Scotland, but instead of being familiar birds, they were birds which we had not previously seen in the wild state. The sun was full upon them, the nearest seventy yards, the farthest a quarter of a mile. The most numerous were brown pintails, the most conspicuous and perhaps next most numerous were chiloe wigeon, and next to them the ruddy ducks. Then came the rosybills and shovelers and only a pair of cinnamon, and we did not definitely identify Chile teal, although I have no doubt that there were many. It is surprising how difficult the teal were to distinguish from the pintails at a distance. The rosybills, of which there cannot have been fewer than two hundred, were a surprise, for Johnson had never seen more than six together before. There were fewer shovelers, but they looked very pretty as they flew and the sun caught their yellow legs as they settled. The large grebes had curiously curved necks which were deep chestnut red. There were various stages of immatures. So there were with the dabchicks, the moorhens, and the coots. The yellow shields of the adult coots were very pretty indeed. The moorhens were grey-sided like crakes. Sitting on some posts at the back were brown-headed gulls and standing on the floating weed was a small flock of waders, most probably lesser yellowlegs from North America.

I began to go over the ducks, bird by bird. In one place, quite close to us, I came upon a male stifftail with quite a different-coloured bill – much greener-grey. It looked a little bit larger and, as it scratched its chin, its bill seemed more spoon-shaped.

I worked on across the flock, and suddenly there it was, as plain as a pikestaff, preening itself – an unmistakable male black-headed duck – the enigmatic, parasitic and euphoniously

Hawaiian geese on Mauna Loa

North American ruddy duck

Mandarin ducks

Flood waters at Threave, whooper swans

A glimpse from the artist's window – pintails coming in

Mallards in a misty morning

Barnacle family

Shelducks alighting on the Dumbles

A pair of teal

Snow geese in the morning

Bahama pintails in the reeds

Domestic difficulties – ruddy ducks

Snow geese after a storm

Siesta

Safe sands for the night

Three pintails

named *Heteronetta atricapilla*. For ten minutes he preened himself in a bustling hurried manner, occasionally swimming briskly for a short distance and then stopping to preen again. The spot at the base of his bill looked yellowish rather than pink, his head was quite black, and his brown flanks were slightly barred. There was a touch of chestnut under his tail. The most important and significant part was his shape, which was quite unexpectedly elongate and in direct contrast to the ruddy ducks round him. He was also buoyant and high in the water, the tail itself appeared to be short, but this is no doubt due to the long upper and lower coverts which form quite a long tail-end to the body. This is comparable with the tail of the brent goose. The shape of the head appeared to be typical of surface-feeding ducks. Indeed, the whole bird appeared to be much more of a dabbler than anything else.

Later, when looking at the fine series of eggs in the Johnson-Goodall collection in Santiago, the smooth texture of the egg surface of *Heteronetta* was very different from the rough surface of the stifftail eggs. These *Heteronetta* eggs were perhaps a little large for the bird, but nothing like so large proportionately as the eggs of the stifftails. They were greenish-buff in colour, whereas the stifftail eggs are conspicuously white. In short, I am now a little doubtful whether the relationship of *Heteronetta* with the stifftail tribe can be maintained with certainty. The little drake we were watching swam over to a patch of weed in which sat five birds. Three of these were clearly ruddy ducks, but two with heads under wings and directly facing us we suspected of being a pair of blackheads. This was confirmed as soon as they woke and our drake joined them. The female was more spotted on the flanks than I had expected, or perhaps more 'blotched' would be a better word. Johnson and Goodall were very much excited about these birds, as it seems that they had never recorded the bird from this lake before, and we also learned that their co-author, Dr Phillipi, had never seen one alive in thirty years of bird-watching in Chile.

It is, of course, remarkable what 12-power binoculars will reveal where 7s and 8s have previously been used, and then the specialist in any bird group has the advantage over the general ornithologist. But I still think we were lucky. Later the same evening Johnson took us to see a flock of about 150 skimmers. We walked a mile along a sandspit on the Pacific shore to the mouth of a small river to see the little flock of these strange birds from the northern hemisphere wheeling aerobatically in the sunset.

SOUTHERN PATAGONIA AND TIERRA DEL FUEGO

We had exceptionally good weather in Southern Patagonia and Tierra del Fuego and saw marvellous birds and filmed most of them. The upland geese were everywhere, but mostly still in small parties from families up to fifty or sixty. Only in a few places did we see greater concentrations. On one oats stubble on the mainland north of Punta Arenas there were six hundred together. The males in this area were almost fifty per cent white and fifty per cent barred whereas in Tierra del Fuego the white birds were well under one per cent – positively rare. On the mainland we saw quite a number of ashyheads up in the hills where apparently ashyheads normally breed. In Tierra del Fuego they were very local, and mainly in the southern part of the island which is more wooded and hilly. We did not see ruddyhead until we got to Tierra del Fuego. Here they were the commonest goose round the farms in the open pampa of the northern part of the island, but there were also some farther south.

We stayed at four places: Punta Arenas, on the mainland in Chile, from which we made excursions northwards; Estancia Caleta Josefina, of 600,000 acres, in the centre of the northern pampa of the island (still in Chile); Estancia Maria Behety, 500,000 acres, farther east, in Argentine Tierra del Fuego, and also farther south, but still in the open rolling country; and

Estancia Viamonte, a smaller one of only 150,000 acres, which is on the Atlantic shore at the edge of the woods and almost in the foothills of the southern mountains. This is the home of the famous families of Bridges and Reynolds who first settled on the shores of the Beagle Channel. At Caleta there were masses of ruddyheads but no ashyheads. At Maria Behety the ashyheads were the commonest goose, but there were some ruddys and the same was true at Viamonte. The geese mix freely, so that the corner of a field next to the farm at Maria Behety which was extra green on the fairly steep slope of the hill had about three hundred geese on it, of which sixty per cent were ashyheads, twenty-five per cent ruddys, and fifteen per cent uplands. In the evening light they looked absolutely marvellous.

An interesting feature is that both ashy and ruddy are most numerous round the farms, whereas the uplands are more evenly distributed out in the country. All the geese are perfectly tame. They are practically never shot, so one can walk within about thirty yards in most places and they hardly get out of the way of a car. After rain there are puddles on the roads and the geese cluster on these, rising a few yards in front as a car approaches. The most important discovery is that I now believe they may not become flightless during moult. We have seen geese in various stages of raggedness, some with several primaries missing – others with complete new wings. Many of the people we have met, mostly English and Scottish, are good naturalists, and the geese are so ubiquitous and common that I simply cannot believe that they could be long flightless without these people knowing it. It raises the point of how much we know about their moult in captivity. Many of the uplands still have young which cannot fly. I have also seen one brood each of black-necked and coscoroba swans unable to fly. Our best sight of black-necked swans was in the Fitzroy Channel north-west of Punta Arenas. Here there were flocks of up to a hundred together, and at some times nearly every bird in the flock was up-ended, looking very droll.

Coscorobas are less common and less gregarious – the most we have seen on one lake was thirty-two all scattered about. They evidently do not normally flock very much. As for ducks, we have had a feast. Crested ducks are without doubt the commonest ducks in Tierra del Fuego. Many of them become largely maritime in winter, but there are still quantities inland at this time of year. The largest flock seen on the seashore was about three hundred, but in many places there were between fifty and a hundred. I cannot see that these ducks are related to the shelducks. They seem to me to have much closer affinities with the dabbling ducks. They are in many ways pintail-like.

The next commonest is probably the brown pintail or perhaps the Chile teal which is everywhere seen in tiny pools and streams, roadside puddles and so on – usually in pairs. I have seen no flocks of these. In one stream we found some flightless Chile teal skulking, and eventually they swam out with no wing feathers – so *they* moult properly anyhow. Chiloe wigeon are the next commonest, almost equal with the pintails and teal, but much more noticeable and therefore probably slightly fewer. They look very decorative in bulk and are rather surprisingly wild. I found one brood of four quite small young and the drake was guarding them as effectively as the female. It is apparently normal, as I saw photos of broods with both birds in attendance.

We saw our first kelp geese on the island of Santa Magdalena in the Magellan Strait. They were a male and two females, but it was not until we got to Viamonte that we became familiar with them. There at high tide they gathered along a short high shingle beach at the mouth of a river. We counted sixty-seven pairs. When the tide went out about a mile, they were spread in the pools and puddles far out among the boulders and the seaweed. The males looked superbly beautiful in flight like white doves in the sun. The females looked surprisingly like large shelducks when seen in flight from behind: there was much more white on them than one would expect. We used to pass this shingle beach each day as we went down the southerly road from Viamonte, so they became quite familiar.

All the kelps appeared to be adults, and we could not understand this until Len Bridges explained that none breed in the area. These were sexually immature birds. They breed on the islands of the south and west coast and the young were expected to appear with their parents any day now for the winter at Viamonte.

But the *pièce de résistance* so far – the bird that has most delighted my eye – is the bronze-winged duck. In Tierra del Fuego it is scarce and local. It lives on fairly fast-moving trout streams, running in rich water meadows and pastures. The streams are just like the Test, where it would be very much at home. I saw my first ones on 15 March when Bobby Huntley, the Manager of Caleta Josefina, led me to a place about fifty miles south-west of his farm where he said he had seen them three years before. We followed the valley of a small winding river and stopped to look across a quarter of a mile of water meadows at some ducks which we could see on a pool. There were brown pintails, Chile teal, and at the next corner three more ducks. One of these was a pintail and it was being chased by one of the other two. As he turned back from the chase I saw the white-faced pattern. He returned to his mate and the pair floated downstream together – the bronze-winged ducks we had come so far especially to see. We started off down the hill to try to get close enough to take a film of the pair. Unfortunately, they swam round the bend so that when we got near they rose with a bunch of pintails and teal. As they flew the female barked like a dog, which is the origin presumably of one of their local names, *pato perro*, the dog duck. As they flew away the whole upper surface of the bird looked black except for the speculum which was bright pale green, but as they swung round to settle it changed to a brilliant light crimson.

They settled on a small pool beside the river and we walked on over a little bridge past some horses until the surface of the pool came into sight. The bronzewings had climbed ashore and stood looking at us, with their orange legs showing up brightly against the peaty heathery bank of the pool. Presently they jumped back into the water and swam to the far end of the pool where the other ducks were assembled – about thirty in all. Most of them were pintails, but there were three chiloe wigeon, a pair or two of Chile teal, and a pair of silver teal. The bronzewings seemed to be the wildest of them all, and hung about ten yards farther away than the rest. However, I managed to stalk up on all fours to within about forty yards and to get some film.

Later when we were staying with Robbie Reynolds at Viamonte we made an expedition to the mountains south of Lago Fagnano, where we hoped to see some of the southern race of the torrent duck. We found no torrent ducks there, but we did find two or three more pairs of bronzewings and, on one small lake, a trio. They were on the far side of the lake and I stalked with infinite patience up behind a small bush to a distance of about sixty yards in good light. Meanwhile, a pair of great grebes had seen me and come over to investigate, giving me good close-ups. Having finished, I stood up from behind my bush and prepared to leave. The three bronzewings stopped up-ending in the shallows, and swam straight across the lake towards me as if I had been a decoy dog. They swam up to within twenty yards and when they began to lose interest I waved my handkerchief, whereupon they immediately came closer still. When they lost interest in that, I squeaked at them and they turned a third time, thus giving me some good opportunities of filming them from about ten yards range. I also had a marvellous opportunity on this occasion and through binoculars on a previous occasion to look at the birds very carefully.

I have always wondered whether they do not have some relationship to the crested duck, but I must admit that they are rather mallard-like in shape, and it may be right to regard them as *Anas*; anyway, it is a most spectacularly handsome animal.

The head and neck markings are very vivid; the male seemed to have a larger face patch than the female. In other respects the sexes are virtually indistinguishable except for a small difference in size. The flank markings are not the rather commonplace mottlings shown in most of the illustrations of the bird, but sharp, black dog-tooth spots, quite large and not very many of them. The back is black and so are the wings above and below – primaries, coverts and all. Underneath there is a whitish patch at the base which may be the axillaries, and which shows in flight, and above of course there is the incredible speculum. So bright is this crimson pink shining patch that the part which is covered by the beige-coloured flank feathers when the bird is at rest shines through it in the sun in a rich pink glow; the orange legs set the whole bird off splendidly. Apart from the name dog duck, they are otherwise known in Tierra del Fuego as wood ducks or black ducks. They are the next largest duck of the region after the steamer ducks – quite substantially larger than the crested duck of Tierra del Fuego.

We then made our way north, to Bolivia, where our objective was to see the torrent duck (*Mergenetta armata*) which lives in the waterfalls of the Andean rivers. For this purpose we travelled over a 16,000-foot pass near the great mountain Huaina Potosi. Near the top of the pass is a tin mine and on the lake just below it we saw a small flock of seventeen Andean geese, swimming far out on the still water. Over the pass we followed a road which leads down the River Zongo.

The River Zongo is one of the sources of the Madeira, a main tributary of the Amazon. It rises in the high Andes of Bolivia about forty miles from La Paz. A hydro-electric scheme on the river supplies electricity to the capital.

A road winds over the pass at a height of 16,000 feet and follows the river in its steep descent down the deep cleft of the valley. At about 9,000 feet the tree-line appears and below this the forest rapidly becomes tropical with tree ferns, a brilliant flora, and hummingbirds. The valley echoes with the roar of the river which is a continuous series of cascades and waterfalls, about fifteen yards wide, the water clear and greenish blue.

The only torrent ducks we found were about two or three hundred feet below the tree-line.

The first view we had of them was from the window of a camionetta (truck). As we curved round the mountainside I thought I caught a glimpse of an upright bird on a stone in mid-torrent. At the next corner we were directly above the spot. Looking down I could see a male and two females sitting on a reddish-brown rock in the middle of the stream. They were about thirty yards almost directly beneath us. As soon as we stopped the car and peered over at them they began to swim downstream with quick nervous head movements.

First impressions were that they were surprisingly like what we had expected. The most striking unexpected feature, however, was the colour of the bill, which is the brightest possible cherry red, with pale yellowish nail in both sexes. The females were very rich chestnut red (pale) on the chin and underparts, the rest dove grey. The males were spotted and streaked with dark brown on a light-grey ground, both above and below, the ground colour on the back being more yellowish. The white head with black lines was most beautiful. In a few moments they had swum out of sight downstream, stopping frequently on stones and swimming down or across waterfalls with the greatest ease.

A little way upstream were two more drakes, standing on a green slippery stone in mid-stream. After a few moments they flew, one after the other, upstream. They did not go far as they failed to reappear at the next corner.

The locals reported more farther upstream, and we made the mistake of driving on in the car about two kilometres, instead of staying with these ducks while the light was fairly good. By walking down beside the river we established that no torrent ducks were present until we returned to the place where the two males were once more standing on their rock, and once more they flew a short distance up. As they took off the individual tail feathers were separated and gave a very 'scrawny' look from behind. The Indian with us volunteered to go back and drive them downstream which he successfully did. One male flew downstream at once, passing below us, and showing its speculum prominently. But for the tail the bird in flight might have been a typical dabbling duck. The second male swam down in front of the Indian – crossing waterfalls where I filmed it. Finally it also flew downstream below us. As we followed we came suddenly upon all five torrent ducks (probably the only ones on this stretch of river).

As soon as they saw us they all hopped into the water and swam about nervously, landing on rocks and then hopping in again. They were about seventy yards away (mainly below us). When they stood on the rocks the head was jerked forwards, the bill above the horizontal and often partly open. A call was just audible above the sound of the waterfalls, and might be described as '*keech*'. Phillips states that no one has previously heard any sound from a torrent duck. While the birds were nervous the call was being repeated almost continuously. It is evidently of a fairly high frequency.

We stood at the edge of the road filming them (although the sun had gone out of the deep valley and the light was very poor). After ten minutes, during which they played in the pool below, leaping out and running up the waterfalls (as mergansers run over the water) and running along the tops of the rocks with an easy gait, the five torrent ducks appeared to be satisfied that no danger threatened, and four of them (two pairs) withdrew to a patch of rapids and began to display. By now the light made ciné-photography with the telephoto lens quite impossible. The fifth torrent duck, a male, sometimes joined in – and may, indeed, have stimulated the display in the first place.

Two notable features of the behaviour occurred frequently. The first was when two males (once all three) raised themselves up by treading water and using the tail as 'surf-board', in order to maintain the body in a near vertical position. The neck was then arched and the bill pointed downward at an angle 50°–60° to the horizontal. The two males were side by side during this display and one usually circled upstream of the other, crossing the current ahead of it, the other revolving so as to keep sideways-on to his rival. I could not establish that the female had any fixed relative position in this manoeuvre. The second movement occurred when the birds were swimming head to current. One would pass ahead of another, and with a quick, rather comical jump, would flick the whole of its hindquarters in front of its rival and throw out a small jet of water into the other bird's face. The water is thrown up in much the same way as it is in the display of the goldeneye. During this display the wing is slightly open and the speculum shows plainly. This was most commonly done by one drake in front of another, but on two occasions I saw a female kick up in front of another female. The displays went on for between thirty and forty minutes, and all this time they were quite oblivious of the observers. From time to time one or other of the party would rest on a rock. Several times a male made his way on to a submerged rock over which the water was rushing. Here he stood with the water halfway up his body. It was amazing how he could stay there with so strong a current sweeping over the rock (which was covered with slippery weed as well).

The females landed on a rock on the far shore several times, but the splashing display went on in their temporary absence.

Sometimes a bird would fly up from the water on to a rock or to the pool above. The take-off was very steep, with no 'pattering' across the water. The landing was made after near-hovering flight. It is quite evident that the wing-loading is as low, if not lower, than in normal dabbling ducks. Since the bird is not 'batwinged', the wings must, in fact, be quite long, but when at rest the very elongated body and long tail give a misleading impression of the wing length.

The plumage of the male in this race is extremely smart.

The black lines on the white head are sharp and extremely handsome. The edge of the streaked plumage of the body is quite sharply defined at the base of the neck. The streaking (and wing speculum) are not unlike those of the mallard. When one male bird was sitting on a rock I thought that without head and tail, the body would pass as a strongly marked rather grey female mallard.

The legs are a dull dark red (in both sexes) with blackish webs, the bill is an amazingly brilliant scarlet lake – less orange than pillarbox red (also in both sexes). The iris of the male appeared black at that range; that of the female was not carefully noted. It was not obtrusive, but may have been lightish brown, for the eyes did not look strikingly dark, as they probably would have done against the pale soft grey of the face if the iris colour had been dark brown or black.

The tail feathers are a notable feature in both sexes. The tail is carried in the water and each feather lies quite separate from its neighbour for the outer half of its length. It is an untidy-looking appendage, but balances the bird gracefully.

On the following day we hoped to be able to film the birds again in improved light, when the tropical sun was almost overhead. But we could only find one female at first. When she saw us she jumped out on to a rock and ran nervously up and down jerking her head forward. After a few minutes she returned to the pool and resumed diving (presumably for food). On returning to the surface she jumped out on to a rock under the overhanging far shore. After ten minutes of very long dives she disappeared and must have used some rock under the near shore and out of sight of us. We did not see her again.

Walking downstream we came upon a pair some eighty yards away. They began to move upstream – mostly walking on the rocks, but often swimming up over the white water. Before I could get more than a short movie shot of them running along a boulder, they disappeared round a corner. Soon after disappearing they must have flown upstream, as three minutes later we were stationed above and below the hidden corner, but, after more than half an hour, our Indian had followed the stream between the two lookout places and no further trace of the torrent ducks was found.

In spite of this disappointment, we had more than our fair share of good luck on the previous day when we had spent so long watching these fascinating and mysterious birds.

From 6th Annual Report of the Wildfowl Trust

26

Saving a Goose from Extinction

The ne-ne or Hawaiian goose is probably the second or third rarest bird in the world. Less than half a dozen ivory-billed woodpeckers are believed to remain in the forests of Florida and Louisiana. Only thirty-two adult Hawaiian geese are known to exist at the present moment, although there may be as many as seventeen more on the slopes of the volcano Mauna Loa. Of the whooping crane of North America it appears that only thirty-seven individuals survive.

A century ago it was estimated that twenty-five thousand ne-nes lived on the island of Hawaii – the only place in the world where the species was found. But the introduction and release of pigs, cats, and dogs, which now run wild over most of the islands, spelled disaster for all Hawaiian ground nesting birds, and the work of extermination was almost completed by the mongoose, brought in to control the previously (and accidentally) introduced rats. At the end of the Second World War drastic steps were needed if this beautiful and interesting goose was to escape extinction; and unlike many threatened local forms of otherwise numerous species, this goose is most distinctive and infinitely worth the effort to save it.

The ne-ne is descended from one of the ancestors of the Canada goose, which must at some remote time have been blown from the North American continent on to Hawaii. Thereafter it evolved very differently from the Canada goose.

With no necessity for migration, the birds' wings have become shorter, and with much reduced aquatic activity the webs of the feet have been partially lost. The body colouring is not unlike that of the Canada goose, but the beautiful buff-coloured neck has developed an extreme example of the 'pleated' arrangement of the feathers found in many other species of geese, while the smart black crown and face and the black ring at the base of the neck are extremely handsome.

In pursuance of one of its main objects, the Severn Wildfowl Trust, soon after its formation in 1946, communicated with the Government of the Territory of Hawaii, enquiring whether steps could not be taken to save the ne-ne, and indicating the world-wide interest in preventing the impending tragedy. But at that time, it seemed, nothing could be done. With so rare a species some form of artificial propagation was clearly desirable, particularly when the species was known to take kindly to captivity. Indeed, a large proportion of the remaining ne-nes were living in a state of semi-domestication in the garden of H. C. Shipman, at Hilo on the east coast of the island of Hawaii.

Three years later, however, the Hawaiian Board of Agriculture established a breeding station at Pohakuloa on the saddle road between Mauna Loa and Mauna Kea, 6,000 feet above sea level, and here a few pairs of ne-nes, most of them lent by Mr Shipman, were established in special aviaries; the project was under the supervision of J. Donald Smith, of the Division of Fish and Game at Honolulu.

Mr Smith was soon in touch with the Severn Wildfowl Trust on various avicultural problems connected with this project, and in December 1949 the Trust's Curator, John Yealland, now Curator of Birds at the London Zoo, went to Hawaii to advise and assist at the first breeding season. Only two of the eggs laid that year were fertile and both the young were successfully reared.

A secondary scheme for saving the species involved sending a pair of geese to England (where the species was successfully bred many times between 1824 and 1900). Mr Shipman presented an adult pair to the Severn Wildfowl Trust and these were brought back to Slimbridge by Mr Yealland. There is no plumage distinction between the male and female Hawaiian goose, and in March 1951 both birds of the 'pair' made nests and laid infertile eggs. It is not altogether common for unmated geese to lay eggs, and that both should have done so in this case was great good fortune, for if only one of them had laid, the cause of the infertility would not have been immediately apparent. As it was, the eggs were removed and a cable was hastily sent to Hawaii. It was hoped that it might be possible to obtain a gander in time for the second clutch of eggs, which are normally laid by waterfowl if the first nest is taken away. Exactly seven days after the dispatch of the cable a fine male Hawaiian goose arrived at Slimbridge, having travelled by air from the opposite side of the globe. Although he took up with one of the females – and indeed, both were very pleased to see him – he was already in full moult (for January is the breeding season in Hawaii); and so the second clutches of eggs were once more infertile.

But in February 1952 both the females nested again and this time the eggs were fertile. The gander, with no more than a faint trace of embarrassment, undertook the care and protection of both females and their nests. As before, the first clutches were taken and placed under bantam foster parents and in less than a fortnight both females had laid again. In all, nineteen eggs were laid of which nine hatched and are thriving; The last five eggs, left to one of the female geese to incubate, failed to hatch, evidently owing to a failure on her part, for three of the four were fertile. The goslings, which are purplish grey with a well defined face pattern, are thriving, although three of them are less strong than the others. The first two to hatch are now becoming feathered, and. like their parents, are quite delightfully tame.

It seems possible, therefore, that by his skill in rearing goslings under foster parents, S. T. Johnstone, the Trust's Curator, will have increased the world population of this rare and lovely bird in one season by more than 20 per cent of its present numbers. This achievement indicates the part which aviculture can play in rescuing a threatened species at the eleventh hour. Such methods might have saved the passenger pigeon, the Labrador duck – even the dodo. The Hawaiian goose is not the only bird to which they should be applied today.

The Times

Postscript

Ne-ne culture, based on the arrival of those first birds at the Wildfowl Trust has, in the sixteen years till 1967, produced 365 young birds (226 at Slimbridge and 139 from pairs distributed to other collections). In the same period the stocks in Hawaii have been increased by about the same numbers, so that the world population now surviving is more than ten times what it was when the first three birds arrived at Slimbridge.

Two ganders bred from a wild gander were sent over from Pohakuloa in 1962 to provide 'fresh blood', and the effect was startling on the fertility and hatchability. Two more have been sent in 1967.

In 1962 the Wildfowl Trust sent its first ne-nes back to Hawaii for release in the Haleakala Crater which is part of the National Park on the island of Maui. Further consignments went in the following three years making a total of ninety-four.

27

Galapagos Expedition

We flew the six hundred miles from Guayaquil to the Galapagos Islands in a Curtis freighter of the Ecuadorian Airline, LIA. The islands are volcanic and lie astride the Equator, but for most of the year they are cooled by the Humboldt Current sweeping up the coast of Chile and Peru from the south. We were there, however, in the rainy season when it is often very hot.

The name Galapagos means 'Tortoise', and the giant tortoises are called *Galapagos* by the Ecuadorians. All the islands have two names – the modern Ecuadorian name and an old English name dating from the days of the pirates and buccaneers. Thus the seat of government is at Wreck Bay on the island of San Cristobal or Chatham.

The largest island, Isabela or Albemarle, is seventy miles long, and altogether nine of the islands are more than five miles long. Only four of them are inhabited, and on three of these there are colonies of several hundred Ecuadorians with a sprinkling of European settlers. The craters of some of the islands rise to five thousand feet and a volcanic eruption took place in late 1959 (after our visit).

The islands show astonishing contrasts: for the most part they consist of dry lava rocks with extensive growth of cactus (*Cereus* and *Opuntia*) which in some places become thirty-foot-high trees. But the higher central parts of the islands are much damper due to the cloud (called *garua*) which hangs over the peaks for much of the year. In some cases the highlands consist of a green moorland but the lower slopes are thickly forested with rich tropical 'moss forest'.

The Galapagos Islands played an important part in Darwin's conception of the theory of evolution. During the voyage of the *Beagle* he noticed that not only were the Galapagos species different from those of the mainland, but that on each island they were slightly different from those of neighbouring islands. This could not be reconciled with the current belief that species were created and immutable. His observations of the finches and mockingbirds and the land reptiles, perhaps more than any other single influence, convinced Darwin of the fact of evolution.

The Galapagos are oceanic islands which have never been connected to the South American continent. Presumably the terrestrial species such as the giant tortoises (*Testudo*), the land iguanas (*Conolophus*), the lava lizard (*Tropidurus*), the snake (*Dromicus*), the gecko (*Phyllodactylus*) and the native rat (*Nesoryzomys*), originally reached the islands by very rare accidents, riding perhaps on rafts of dead timber and weed which came drifting down the mainland rivers. One such accident might bring more than one species at a time but it seems certain that there have been several separate 'arrivals' of this kind. At one time the main part of the archipelago was probably a single much larger island upon which the initial colonisation took place. Subsequent subsidences left only the tops of the mountains as separate islands, on which the various species developed racial and even specific characters. A number of sea level changes of this nature in both directions may have played an important part in the startling adaptive radiation shown by Darwin's finches. On the island of Isabela there are five volcanic peaks and five different races of giant tortoises have been reported from this one island (all are believed to be either extinct or on the edge of extinction). On this island the marks of marine molluscs (*Chiton*) and echinoderms have been found at more that three hundred feet above present sea level. Clearly there have been many changes in sea level since these islands first appeared.

Darwin described the fauna of Galapagos as eminently curious, and this it most certainly is. The native animals are almost all unique to the islands even if only as in some cases a Galapagos subspecies. A number of them are still probably as numerous as they have ever been but others are seriously threatened by the direct and indirect effects of human colonisation. A good example of the difficulties which they face is given by the tortoises already extinct on several of the islands and clinging on by the narrowest margin on others. On the island of Santa Cruz or Indefatigable we were lucky enough to see seven of these magnificent reptiles, the largest about four feet six inches long. They are hunted for their flesh and for their oil by the Ecuadorians; their eggs are dug up by the hordes of feral pigs which have run wild on most of the islands; their young, while their shells are still soft, are eaten by the feral dogs; and their green food is eaten (in some places bare) by feral goats, donkeys, and cattle.

Other species whose numbers are dangerously small are the flightless cormorant – the largest cormorant in the world – and the little Galapagos penguin; it is doubtful whether more than a few hundred of either exist. Two more very rare species are the flamingo, which may be no different from the West Indian flamingo (though it may prove to be distinct), and the fur seal, which is confined to a very small number of colonies. The land iguana, a splendid nearly three-foot-long brown lizard with yellow head and legs, is now to be found only on four islands.

Happily the marine iguanas, the most striking of all, are still common on most of the islands. Although the races have not yet all been described, the populations are clearly distinguishable on each island. The iguanas of Tower Island in the north are small and pitch black for example, while those of Hood and Floreana are rose-pink with black spots, a green crest and green front legs. The race which lives on Fernandina or Narborough is brownish black and has the habit of collecting together in very large colonies; we found nine hundred of them in a solid mass on the top of a rock about twenty yards long at Punta Espinosa at high water, with the scarlet and blue shore crabs climbing all round and over them. At low tide these iguanas go out to forage on seaweed up to ten and possibly twenty feet down on the outer reef. We found that they normally lay only two eggs, buried in the sand at the tops of the beaches. We filmed the females digging their burrows and actually laying the eggs – with an eight-minute pause between first and second. The hole is then filled in and covered over by the female with great care.

The Galapagos sea lion is still quite common; we swam with them and filmed them underwater, and were twice chased out by the very aggressive bulls. The Galapagos dove, a beautiful little bird which is rather disastrously tame, is numerous only on some of the uninhabited islands, having been greatly reduced on the inhabited ones. The hawk and the herons of three species seem to be holding their own, as indeed are the finches and mockingbirds. Unfortunately, we did not see the famous woodpecker finch, which uses a cactus spine to pry insects out of cracks in bark just as a woodpecker uses its bill.

The lava lizard and the small non-poisonous snake, both of which seem to vary from island to island, do not appear to be in any danger of extermination, though the snake is less common than the lizards: but the interesting native rats are unable to compete with the introduced black and brown rats, and may well be on the way out. Breeding stocks should if possible be established in captivity.

The status of the small dusky Galapagos albatross, which breeds on Hood Island and nowhere else in the world is rather obscure. None were present during the period of our visit, though it is understood that many had returned only a month later. Hood is rarely visited but the entire species appeared to be represented by one or two hundred individuals.

Our base of operations in Galapagos was at Academy Bay on the island of Santa Cruz or Indefatigable. From here we made two ten-day voyages to the other islands in the patrol boat kindly put at our disposal by the Ecuadorian Navy. Altogether, we landed on eleven of the islands in the group.

Although living conditions during our trip never descended to the level of hardship, there were many times when they could have been described as 'rugged'. Fresh water and food are both scarce, and many nights we slept out in spite of a minor mosquito nuisance.

We swam frequently among magnificent fish shoals, marine iguanas, sea lions, and penguins, and were not unduly troubled by sharks – which are quite numerous. The sea lion bulls were reputed to be the most dangerous animals we were likely to meet, but although they looked aggressive they never actually attacked in the water, and one which did so on land came to within a few feet as I filmed him and then stopped. It was only when the film shot was ended and I looked up from the view-finder of the camera that I realised he had been charging me.

Some of our best film was made of the ecstatic courtship of the seven-foot wing-span frigate birds on Tower and Hood.

On Wednesday 11 February we landed in a small cove at the back of Darwin Bay, Tower Island. Darwin Bay is a remarkable natural anchorage – a crater is tipped sideways so that ships may enter over one lip into a circular harbour a mile in diameter. The cliffs overlooking the cove in which we landed were painted with the names of visiting yachts and the years of their visits. Rough white lettering was daubed across all the accessible vertical rock face, which at first sight seemed to us to be inexcusable vandalism; but later when we thought about it more carefully we came to a more tolerant conclusion. The Galapagos Islands are remote and for any but a naturalist uninhabited Tower must be a forbidding and desolate place. The loneliness was lessened by the untidy white names which peopled the anchorage in imagination.

There was also, as we landed, a stink of fish on the beach from decaying fish-heads left by a fisherman's gutting party some months before. The shallow cove was full of large sharks, and walking on the white coralline beach were a few rather small quite black marine iguanas. The rocks on one side were red with shore-crabs, and in the small bushes above the high tide mark there were nesting frigate birds, the males with their huge red throat sacs inflated, already courting, while the young of the previous year with whitish heads still hung about near their old nests.

Dimorphic red-footed boobies, some white some brown, but always with blue bills and red feet, sat about in the stunted shrubs; blue-faced boobies with large young looked almost exactly like our northern gannet. But the most graceful of the sea birds there, with a beautiful sad evocative cry, was the Galapagos fork-tailed gull, with a sooty black head, a sharp white spot behind the bill, almost like a drake goldeneye, scarlet eyelids, and an orange red gape – altogether a strikingly handsome gull. The quite common dusky lava gull was also on the beach at Tower. Although much less beautiful it is of a particular interest because its nest has not so far been found. It may breed high on the mountain tops.

But in the pool behind the beach was the most exciting thing, our first pair of Galapagos pintails. They are much as I had expected, except that the female is duller; there was no red on the bill of this one and very little white on the cheek. In fact, she was more like a grey teal than a Bahama pintail. They were quite tame, and we finally filmed the pair at twenty-five feet in the clear water of the tidal pool which had filled by filtering through the coralline beach. Also on this pool were a family of yellow-crowned night herons and a small blue reef heron. The Tower Island mockingbirds hopped round our feet and on to the camera tripod, and among the opuntia trees were two species of black finches, one with a very large bill and the other with a much smaller one. On the following morning there were thirty pintails on this pool, and I established that yesterday's female was duller than most. Several females had quite prominent red spots. Later when we climbed to the crater we saw a lot more, so that there might have been fifty or sixty altogether on Tower.

On a lagoon behind the beach at James Bay on the island of James or Santiago, there was a little group of twenty Galapagos pintails near a flock of twenty-one surprisingly wild flamingos, and we saw some display which to me was indistinguishable from that of the Bahama pintail. We met a pair later up in the tortoise country in the interior of Santa Cruz, where they were sitting on a little pool among the rain sodden woods; a few yards away a giant tortoise was feeding. I saw one more pair on Hood, in a puddle on the path on the way up to the old war-time radar station now deserted and in ruins in the centre of the island. The Galapagos pintail is perhaps a little smaller and more slender, and if anything more graceful, than the Bahama, but it is less smartly coloured; the bills of the females seem usually to show less red, and the sharp line dividing the white cheek and the brown crown in the Bahama pintail is much softened and blurred in the Galapagos race. In overall colour the bird may be a little reddish brown, though in the field there does not seem to be very much difference.

From 11th Annual Report of the Wildfowl Trust

28

September in the Country

To me September has one particular and special significance, because it's the month when the wild geese first arrive in Great Britain from their breeding grounds in the far north. And of course, war or no war, they'll come just the same – and year after year they arrive with *extraordinary* regularity. Today, tomorrow, and the next day, the 24, 25, and 26 of September the main flocks will be collecting, as they do without fail every year, although odd small parties may have been seen about for a week or two.

Well, we may think that our country is a very different place from what it was last April when the geese went off to breed, and indeed, for human beings there could scarcely be a bigger difference cast over the face of the world; but for wild geese on migration there will only be one real difference that they can detect at once – the black-out. At the end of their long flight across the sea they won't see the twinkling lights below, and if a wild goose ever connected a light with his arch enemy, man, he might well wonder what calamity had befallen the human race that the earth's surface should be in such abysmal darkness. Actually, I don't suppose any wild animals ever really think of a light as something man-made. It is probably to them just another kind of animal that twinkles, just as an aeroplane is another kind of bird that roars.

But I am wondering whether this black-out will really make

any difference to migrating birds, because, although of course birds migrated across these islands long before even a cave man's fire twinkled below them, yet within the memory of any *individual* bird which is now migrating, there have always been lighthouses and the glare of towns and so on, as extra landmarks in the night.

Ornithologists have practically no satisfactory theory to account for the extraordinary direction finding abilities of migratory birds. We can only suppose that they must be equipped with a sense which is either altogether lacking or completely undeveloped in us.

Some clues to this sense have been discovered by experimenting with homing pigeons, and in one case it is said to have been established that pigeons let out near a large wireless station took much longer to find their direction and set off, when the station was actually transmitting, than when it wasn't. Well, whether they have the equivalent of a compass inside their heads or whether they use quite different principles to find their way, I'm quite sure the geese *will* find their way to their marshes and mudflats, black-out or no black-out.

Most years I manage to get to Scotland during this week in September because that is where they first arrive and I like to be there to meet them, but this year – well there have been other and more important things to do. But I had a telegram yesterday to say that they had begun to arrive and I know that for the next three or four days they will be pouring out of the sky down on to the great sandbank on that estuary that I know so well. They're called pink-footed geese and more than any other kind they are *our* geese, because most of the pinkfeet in the world spend the winter in England and Scotland.

They arrive in small bunches of fives and sixes, up to twenty and occasionally sixty or a hundred, and if you sit, as I've so often sat, where the green saltmarsh breaks down to the sand, scanning the northern sky, you'll hear a distant call and then you'll pick out maybe a little line of dots, high up at first but gliding down and swinging round to settle amongst the great mass that is already collected on the high sand.

When the last have arrived, there will be ten thousand geese on the estuary, and when they rise there will be a thunderous roar that you could hear three miles away. That first arrival of the geese has always been one of the greatest thrills of the year for me and really I don't believe *anyone* could watch it unmoved. But of course the places where one can see wild geese arriving on migration are few and far between, though one may often enough see a little skein of them – a V formation – passing high overhead. It's very hard to tell what kind they are except by their voices, which are quite characteristic once you know them. A friend of mine was fighting in Spain and he told me that right in the middle of the battle of Guadalajara a little V of geese flew over and he would have me believe that he held up the whole battle whilst he stopped to watch them with his field glasses, and to listen to their cry so that he could tell what kind they were.

During the last few weeks many thousands of wild geese have been migrating over battlefields because one of the biggest autumn migration lines of waterfowl comes through the Polish marshes leading on south through Hungary to the Mediterranean. The Vs of geese and the Vs of bombers must have passed each other in mid-air.

I dare say that it'll be some time before I see the first wild geese of *this* winter, but you know I think it's rather good for our sense of proportion to see how little nature is upset by the doings of mankind, and so whenever I *do* get into the country, I find there's not a moment to waste.

My first thought on a bright September morning is: 'It's perfectly impossible that there can be such a thing as a war when the world is so beautiful,' and my next is: 'that for some reason the world, the fields, and the trees and the rivers are so much more beautiful because there is a war,' and then I am ready to plunge into the details of birds and beasts and plants without feeling any more that they are irrelevant. Then I can spend half an hour looking for a poplar hawk moth caterpillar on a little poplar sapling, decide that he must have just gone to ground to turn into a chrysalis, in spite of the freshly eaten leaves, and at last just as I am turning to go, spot the great fat green fellow masquerading as a young leaf right under my nose.

Caterpillars are masters of camouflage, and if it were possible to conceal a munitions factory half so well as the caterpillar conceals his portly person we shouldn't have much to fear from fast bombers. Of course, it is possible to do quite a lot to merge a big block of buildings into the surrounding countryside by the addition of some tons of various coloured paint, and the man who says where each colour shall go is called a camoufleur, which I always think of as a sort of cross between a camel and a cauliflower. Anyway, that's very often what the final patterns look like – at any rate, when I make them; because you see, being one of the many who must wait for the telegram which calls them up, my interim job last week has been trying to make a hundred acres of factory look like a wood with a couple of cottages in it.

But in this fine month of September I did manage to get away into the country once, and that was when I went to see a friend who lives on an island on the east coast. It is a flat, marshy island in Essex, and there is a road leading to it which is only uncovered at low tide. When I arrived I didn't know the times of the tides and I drove over the sea-wall to be confronted with the sea and the road ahead disappearing into it. Out in the estuary about a mile away was the island – flat and surrounded by a sea-wall – with a low undulating hill in the middle, and at one side a clump of trees with the farmhouse nestling amongst them.

I had to wait about an hour for the tide to fall off the road enough to let me proceed, and even then I went slowly, waiting for the water to run off the deeper places.

During that time big cumulus clouds blew across the landscape so that in shadow the distant sea-walls looked black, and in sunlight brilliant green like the sea-grass on the mud when the tide uncovered it. This sea-grass is the food of the little brent geese which come there in winter, but of course there were no brents there then. But there were some ducks – about a dozen mallard left high and dry asleep on the mud, all of them in dull brown plumage, because few mallard drakes get their full winter colours until the very end of September.

I watched a black-backed gull eating a large crab on the mud within thirty yards of me. The crab was alive to begin with but it never had a chance to use its pincers because the gull was too quick with lightning stabs of its bill, and at last only the claws and the shell were left of the poor old shore crab.

There were lots of waders on the shore, mostly redshanks, but also a good many dunlins, tiny little birds whose legs twinkle so fast as they run over the mud that they seem to be rolling like a little ball. Some of them were still showing the black tummies of their summer plumage and amongst them was a very smart black-breasted bird, a grey plover, still in *his* summer plumage, too. I think a grey plover is almost the most handsome of all our wading birds with his jet-black breast and face and his pearly grey back, and when they do come they are almost always in ones and twos and almost always amongst dunlins.

I splashed along in my car through the puddles and ploughed through the shallow fords of the creeks, sending up showers of salt spray which no doubt will rust my car to bits (for it looks as though the poor old thing will sit in the garage for the duration, getting redder and redder!)

On either side the level of the soft mud was four feet above the road and the dunlins ran about on it quite unconcerned with the approach of the car – to them, no doubt, another blundering animal which makes a nasty whirring noise.

When I got to the island I saw a pair of short-eared owls hawking along the marsh, and three or four curlews flew from the meadow out on to the mud.

I arrived at the farm to find that my friend, who spends most of his time exploring in Greenland, was building a chicken-house, whilst his wife was busy de-maggotting sheep with some very strong-smelling disinfectant. In the farmyard fifty swallows were gathered on the wireless aerial, packed together, just ready to be off, for as the geese arrive from the north the last of the summer migrants leave for the south.

After tea a hundred day-old chicks arrived and we put them in the now completed brooder. Taking them out of their boxes it was quite incredible to think that these lively little fellows had never eaten anything in their short lives. They had kept each other warm, twenty in a box, during their long train journey, and they were as bright and perky as if they had just been taken from the hen that hatched them. We sprinkled meal on them and they started to pick it from their neighbours' backs, and so tasted their first food.

I could tell you lots more about that farm and that beautiful island among the mudflats, but I mustn't steal the subject from the owner, who quite often broadcasts and may want to talk about it himself one day!

As I came away across the causeway, just before the tide covered it, I put up a big flock of teal – about a hundred of them which had been sitting on the mud. They swept down the estuary flying almost like starlings and behind them came five wigeon – the advance guard of the great army of wigeon which will spend the winter there.

The arrival of wigeon in most parts of the country is the real herald of autumn. A few wigeon breed in northern England and Scotland, but only a mere handful by comparison with hundreds of thousands which come to our coasts in winter.

This winter they and the geese will most probably have a more peaceful time than usual. There will be fewer humans who have time to pursue them.

And they aren't the only birds that will benefit from lack of human attention. There are some of the rarer hawks too, who may get an extra lease of life when the preservation of game birds ceases to receive so much attention.

September is three-quarters over and the autumn has arrived. Autumn that may seem to symbolise sad things – things that are coming to an end, but *I* always like to think of it as ushering in new beauties, and bringing with it new magic, on the wings of the wild geese.

From BBC Home Service broadcast, September 1939

PART II

NAVAL OFFICER

29

Nature of Fear

In the summer of 1940, German bombers had command of the air. On a hot afternoon in the English Channel they came quite suddenly out of the dazzling haze round the sun. A spatter of machine-gun bullets was the first warning; tracers drew thin blue hairlines round the destroyer; and then with a whining roar the first of the Ju 87s pulled out of its dive.

Its bomb fell in the sea fifty yards away and a small spout of white water rose in the air to be engulfed instantly in the vast brown upheaval of the explosion.

Huge pinnacles of water shot upwards as the destroyer reeled and shook.

Ordinary Seaman McGill was knocked down by the blast and the jolt, and then nearly washed overboard by the deluge of water which fell like a cloudburst on the upper deck.

Dim through the brown mist of it he could see the next Stuka steep in its dive. Look out now, he thought, here's another. Quick now.

He scrambled up and scuttled for the protective cover of the galley flat.

The whine of the power dive was followed by a shattering explosion. The whole ship lurched and shook from end to end. The sound was as if the largest tin in the world had been whacked with a gigantic stick. With it came the roll of thunder and the crack of doom.

McGill was certain that, in this his first experience of action the ship had been hit, that she must sink instantly; but outside the guns were firing still.

He had been on the way to his Action Station when the first bomb fell, but now he was back in the galley flat with the roar of aircraft and the bark of pom-poms in his ears and the fear of God in his heart. He crouched in a dark corner, the animal instinct to hide in complete control of his body.

The main armament had opened fire; the blast of the 4.7 guns blew into the flat through the ammunition hatches, and the crash of their discharge added to the shattering din. The ship heeled steeply under full rudder at high speed. Supply parties were handing up ammunition through the hatches; repair parties stood waiting.

No one noticed young McGill in the corner by the galley door. He crouched there with both clenched fists held against his head: his face was screwed up, lips apart, eyes half closed. His pulses raced but his mind was numb.

He knew there was something he should be doing but he could not think what it was. Could not think . . . If only there was a chance to think, but everything was noise – noise above anything he had ever imagined – noise that was more felt than heard, that shattered and stunned him.

From very far away he could hear the voices round him.

'Come on – keep that cordite going!'

'Near miss – lots of little holes in No. 2 Boiler Room.'

'More boxes of twelve-pounder, quickly!'

'. . . crashed about half a mile away.'

'Poor old Lofty.'

All these things he heard clearly, though far away as in a dream; they did not belong to his world at all.

Suddenly the noise stopped. Someone shouted, 'First Aid Party – starboard waist,' and three hands shuffled past him with a stretcher.

McGill remained in his dark corner. Still he could not move. He was shivering violently. I must go . . . I must . . .

It was the final effort of his breaking will. He felt he was floating out towards the sunshine from the dark imprisonment of his terror. He was dazzled by the bright sunlight, but with it strength and self-control returned.

He met the stretcher party coming in and recognised the man they were carrying. It was old Hawkshaw, who had been teaching him boatwork from a book that morning. When you had grown up with boats and knew as much about them as McGill had learnt at home in Ireland, what with fowling in winter and fishing in summer – all these special names out of books were very bewildering. And what had books to do with this horror that he saw?

Old Hawkshaw! His face grey and twisted; the leg of his overall torn and bloodsoaked; a long dagger of broken bone sticking from the mangled flesh; the hand that had held the book that very morning was not there now.

The stretcher passed, and McGill went out into the afternoon sun.

At the foot of the foc's'le ladder was a great pool of blood, and lying by the funnel was a headless body which someone had partly covered with a hastily thrown duffle coat. The sea-boots projected grotesquely. There was movement under the coat.

'Poor old Lofty Reynolds,' said a voice at McGill's side.

'But he's still alive,' said McGill. 'I saw'm move, so I did.'

'Don't you believe it – that's just nerves.'

McGill went on up to the foc's'le and sat on a locker near the gun. The bright sun shone down on him. He felt very sick. The rest of the gun's crew were far too busy to notice him.

In the heat of the action the Captain of the gun had never missed one of his extra loading numbers, so no one but McGill himself knew that he had spent those four thunderous minutes crouching in the cover of the galley flat instead of feeding cordite to his gun.

The enemy did not return that afternoon, and when the

protecting Spitfires arrived the guns' crews fell out. Most of them sprawled about the upper deck in the sun in sudden relaxation.

McGill was stretched along the length of the locker gazing up into the deep blue of the afternoon sky with half-closed eyes. He was dazzled by the bright sunlight. His breath still came in choking sobs but the numbness was passing. As with cold hands, a terrible smarting pain took its place, filling every corner of his mind.

What difference did it make that no one else knew he had done it? *He* knew. Yes, he knew, but he did not understand. He had depended so much on being brave; he had been so certain that he would be.

It wasn't as though he had never seen fear before in all his eighteen years. They had met face to face in the short steep seas between his home and Chapel Island when he tried to cross on the ebb with a southerly gale.

Those were the nights when the wild geese flighted low and he loved to take the boat across, although they said he was mad to go. If it was dangerous he liked it more.

He would come back long after dark, soaked through, with perhaps a goose and a couple of wigeon to show for it, his hair wild and tangled with the rain and the wind, and his cheeks flushed, and Tim, his brother Tim, would give him a look ... like that ...

He had beaten fear many times, many fears: the fear of strong winds and strong tides, and fogs and quicksands, and leaky boats. The lough found new ones, but he had their measure – on the soft mud flats of the lough where others were scared to go.

The lough. And Chapel Island.

Just to think of them!

And the bright pastures of Green Island across the soft stillness of the bay; the golden brown of the seaweed and the smell of it; the sighing of the ash trees round the house, and the gentle curve of the grey pebble bank which joined their own island with the mainland shore, and was only covered by the highest spring tides. These were the real things, more real a million times than Ju 87s and destroyers and HETF shells.

The green mound of the island with its tiny farmhouse among the trees on its sheltered side was his home and his life. He longed bitterly and desperately to be there, never to have left it, to wake up from the nightmare that was today.

There was aching pain in remembering his Irish home, for the last passionate sight of it, the whole last scene of departure was the clearest memory of all.

It was not anything they had said. His mother had been very quiet. He had loved her especially for that. 'It'll not be long, boy,' and she had smiled.

And then, just as his father took up the reins and the cart that was to drop him at the bus stop began to trundle out of the yard, he had looked down at Tim standing below. Tim, not quite two years younger than himself, who had shared every adventure until now.

He had looked very deep into Tim's grey eyes just for a second. In that second he knew that Tim was giving – giving in an agony of generosity all he could of his strength and love, an essence of his spirit. Take *my* sword, too – you may need it.

McGill, by meeting Tim's gaze, had accepted his trust – and now he had betrayed it in his first moments of fear.

Tim would make excuses for him if he knew.

Tim would say, 'But you couldn't help it in all that noise. That wasn't your fault.'

But it was. He was a coward.

If only he could see Tim and talk to him. It would ease the shame . . .

'Don't you want no supper, lad?'

McGill opened his eyes and saw the burly figure of Ordinary Seaman Money standing over him.

'No,' he said.

'They piped "hands to supper" long enough ago.'

'I don't care,' McGill said fiercely.

'Ee, what's t'matter wi' you?'

'Oh, go away.'

Copper Money scratched his ginger head. Most people were pretty talkative this evening, and it wasn't as though McGill had been a special pal of Lofty Reynolds, though of course that might be it. But you never knew where you were with young McGill anyway – a proper Irishman he was, with his black hair and pale blue eyes. Copper didn't know much about Irishmen, but it was a convenient explanation.

He lit a cigarette and sat down on the deck with his back to the foc's'le screen. The sun was setting into a soft grey cloud which lipped the edge of the glass-calm sea. High streaks of mackerel sky turned from silver to gold and from gold to dull red like hot iron cooling.

Copper puffed at his cigarette. Queer chap, young McGill. You never quite knew what he would say next. Then sometimes he had that faraway look in his eyes as though he didn't see or hear what was going on about him. Those were the times when the Petty Officers used to get mad at him for not obeying orders quickly.

Copper had once heard the First Lieutenant talking about him. 'What, *that* good-for-nothing? He can't even salute properly.'

Copper knew this was an injustice. Although the boy was long-limbed and clumsy there were things he could do – chucking a heaving-line, for instance. He just had the swing of it. And he was a marvel in the whaler, quick as a cat for all his long legs, and he knew how to pull and all.

Besides, he was somehow different.

Copper remembered the night he had first discovered it, on watch at the twelve-pounder. They had been looking towards the moon and its long sparkling lane reflected on the calm sea. McGill had begun to talk about his home. It was the first time Copper had ever heard him talk like this – even his voice seemed different, suddenly alive.

He had talked about the moonlight across Strangford Lough and the lapping of the flood tide over the soft mudflats and the stillness broken by the calls of the various birds – he knew the different ones too. Different names they all had. And he had talked a lot about the beauty of it, as though he'd taken special notice of it, like one of those artist chaps.

But he had liked most to talk about the wild geese – the barnacle he had called them – and the music they made when they fed right into the bay below the house. 'Oh, it was rare music, so it was.' There was a dreamy quality in his voice, soft like the rain, but with a secret urgency.

Copper, sitting on the Ready Use Locker with his legs dangling, had seen a vision. Something from a different entrancing world, something that you never knew about if you lived in a grey back street in Warrington. He listened happily, and when the watch was relieved and the spell broken he felt sort of flat.

Oh ah, he was a queer one, was young McGill. There he lay now on the locker, brooding, and thinking maybe of the first casualties he had seen – nasty ones to see too – of poor old Lofty perhaps. Who could tell what he was thinking?

As Copper looked at him, he saw the fingers of his hand curl into a clenched fist.

Three months later, on a wild November night, McGill was keeping the middle watch as a lookout on the bridge of the destroyer, now engaged in convoy escort duties in the North Atlantic. His eyes, long practised in detecting the dim blackness of a pack of wigeon or a party of brents on a dark night, easily picked up the row of black lumps which were the ships in the starboard wing column of the convoy.

It was very cold. The gale howled in the foremast rigging and spray burst from time to time over the bridge, rapping viciously on the screen and on the sou'wester tops of the men who ducked to avoid the full lash of it.

Suddenly there was a dull explosion.

'See anything, port lookout?' asked the Officer of the Watch.

'No, sir.' Thick scurry of rain and sleet obliterated even the black lumps of the convoy.

'You heard an explosion, didn't you?'

'Yes, sir,' said McGill.

The Officer of the Watch moved to the voice pipe. 'Captain, sir.'

A sleepy voice answered him.

'Explosion, sir. I think it was a torpedo.'

The Captain reached the bridge in time to see the rocket signals, dim and blurred through the rain, which confirmed that a U-boat had attacked the convoy. He thought, What a night for it, with this sea and these squalls and wind. The U-boat's bound to get away with it. The enemy was developing surface attack at night in these early days of the Battle of the Atlantic, when the issue lay in the balance.

A hunt was carried out and many depth charges were dropped, with unknown result. McGill heard and felt their explosion almost unmoved. He was at his Action Station now at A gun. Great seas washed constantly over the deck, but he crouched to leeward of the foc's'le screen trying to keep dry. At any moment a torpedo might be running towards them – at this moment it might have started . . . or at this . . . He was not at all afraid.

The air was full of salt mist.

About an hour after the torpedo struck the Captain received orders from the Senior Officer of the escort to stand by the sinking ship. She was on fire.

McGill could see her through the rain as a big nebulous ball of orange fire. Below this the haze of spume was thick over the angry wave tops.

The destroyer crept up to leeward of the wreck and lay no more than half a cable away, rolling wildly in the trough of the sea. Her ten-inch searchlight swept the intervening blackness and lit upon a boat, low in the water and packed with men. It looked stark and white in the beam, the faces shining up into the light through the mist.

McGill heard the thin pipe of the Bosun's Mate above the howling of the wind, followed by a shouted order, 'A and B guns' crews muster in the port waist.'

That meant him, that meant a job to do. Up till now it had all seemed far away, someone else's responsibility in which he could only stand and watch, shivering with cold and a vague detached excitement. Now there would be work for him, and he suddenly realised the strength of his burning longing to help – to help those blank white faces clustered together in fear – to get them at all costs, every one of them.

There was a crowd round the fo'csle ladder, filing down to the port waist, and when he reached the iron deck the boat was nearer, but it was also lower in the water – it was almost waterlogged. The wave tops were slopping over the gunwale.

McGill heard the sharp crack of a rifle shot, and a thin white line appeared in the searchlight's beam. The line-throwing gun had been fired over the boat, the thin cord flying out in a low arc. But the wind blew it away like gossamer, far astern.

McGill looked up to the pom-pom deck, where the Gunner's Mate was reloading the rifle. Presently he saw the figure poised, ready to fire again; then came the crack and a second line sped out far ahead of the boat to make allowance for the wind. For a moment it hung in the air, then slipped sideways in the clutch of the gale.

The men in the boat reached upwards, lurching, but it passed over their heads.

'Heaving-lines!' shouted the Petty Officer, who was in charge of the party waiting on the iron deck. 'Hey you, McGill, get a heaving-line, chop-chop!'

Swinging from one handhold to the next, McGill made for the foc's'le where the heaving-lines were kept in the foc's'le caboose. He knocked off the clips and opened the door, closing it quickly just before the next wave broke over the foc's'le.

In less than a minute he was back on the iron deck with a line, the specially weighted one with a brass nut inside the turk's head at its end.

The boat was much closer now. He gauged the distance and thought: I can just throw it that far. All the time he was coiling the line on his left arm, ready to heave it.

When he was ready the Petty Officer said, 'Don't throw yet,' but McGill had gone far away into a world where his own capabilities were finely drawn like silver point. Far away and far ahead too.

He could see the men in the boat reaching up to catch his line; he could see them hauling over the heavier grass line that was now being flaked along the deck behind him; he could see the boat alongside and see the men being rescued, hauled up to safety.

It would all be done – if he could throw it that far. Just that far.

He swung back, and with a great bowling action cast the coils out, aiming just ahead of the boat. The line spread itself smoothly, and the weighted end, falling like a trout fly, lipped the gunwale.

A strange murmur, half heard above the roar of the storm, ran through the cluster of men at the guard-rail. Now they could help. 'Ah – now then,' they said.

The grass rope's end was bent to the heaving-line and passed outboard as the men in the waterlogged boat hauled in. At last it reached them and was secured.

McGill watched the boat being dragged towards the destroyer in angry jerks. Every time the ship rolled away the grass rope came taut, and when it rolled towards her they took down the slack and held on for the jerk. It was an unseamanlike method, but the First Lieutenant was giving the orders.

McGill could only gaze at the scene with fierce intensity, every muscle braced, ready to move when the opportunity came, to do anything he was told, or anything he saw that needed doing. He had never been so ready in his life, not even on those wild nights landing on the rocks of Chapel Island, or driving up the lough with the tiny sail which overcanvassed the duck-punt to the point of breathless danger.

It was raining harder now, and with the squall the wind was whipped to new fury. The boat and the crowd on the tumbling destroyer's deck and the thick atmosphere of rain and spray were all lit by the pinkish-orange glow of the burning ship in the background.

Amidships she was a furnace, high flames leaping from the bridge and flickering from the ventilator cowls. The plating of the superstructure was already red-hot. The hot pungent smell of burning blew down across the destroyer's decks in occasional waves: a stifling, menacing smell.

The crowded boat was nearly alongside now, heavy and sluggish in the great seas. About twenty men sat upright on the thwarts, and lines were being passed to them to haul her along side the special clambering net which hung down the destroyer's side. Faces below were upturned, arms reaching already for the net, some of the men crouching, some standing.

McGill saw other figures in the boat down in the water, floating – swilling sickeningly back and forth with the rolling of the boat. Why didn't they pull them out of the water, give them a chance?

With a splintering crack the boat came alongside. A dozen jumped for the net and hung on. The destroyer rolled away and they were dangling against the ship's side far above the boat. Then she rolled back; the water came up to meet them, and in a moment they were in it. The boat was caught by a wave and swept again towards the ship's side.

McGill saw what would happen, saw it helplessly, agonised. He shouted, 'Look out, now,' like most of the others watching, but it was too late.

The men in the water also saw, saw the boat high above them on the crest of the wave.

A terrible shout of despair that was almost a scream came up to the watchers and was blown away into the wild clamour of the night.

Then the boat crashed against the ship's side for the second time – a muffled crash as the human fenders took the blow.

At that instant McGill suddenly moved, his heart beating high, his brain crystal clear, his purpose defined. If he had felt fear at all that night, no trace remained to cross his mind. There were men in the water, helpless, injured, drowning; men who were within reach of safety but slipping away. It was for this he had been waiting – for just such a chance.

He took off his duffle coat and climbed through the guard-rails and down the clambering net, past the survivors who were struggling upwards. A moment later he was on the gunwale of the boat which still bumped and crashed against the ship's side.

Floating near the stern was a limp body. He seized hold of the kapok lifejacket and began to heave the man back into the boat.

In a moment he was being helped, and he heard the rich north-country voice of Ordinary Seaman Money in his ear. 'That's the stuff, lad. In with 'im!'

'Get the other chap,' McGill said. He hauled the heavy man into the boat and propped him against the thwart, then turned and helped Copper with the second.

He had never been so strong, he could do anything. It was dangerous the way the boat crashed against the ship – oh yes, it was dangerous, but he was nimble enough to avoid the danger if he kept on his toes. The boat would break up soon so they must hurry, but he could be quick and strong, never so strong.

With a mighty heave the second man slithered into the boat.

Two more hands were on the clambering net now, and coming down to help. Heaving-lines were passed and bow-lines made round the bodies. Besides the crushed there were the three floating in the bottom of the boat who might yet be saved by artificial respiration. As they tied the bowlines, and struggled to lift the heavy burdens, willing hands reached down from the deck when the boat came up on a crest and dragged them on board.

'Below there,' came the First Lieutenant's voice. 'There's a fellow floating just clear of your stern there – can you get him?'

McGill saw him in the light of a torch from above, about five yards away, a white lifeless face above the lifejacket, almost as much under the water as above it. But there was a chance!

He slipped over the side and struck out for him.

It was breathtakingly cold, even though he had been wet through before, but it was easy now, no effort to try out his strength. His breath came in sharp gasps, the wavelets slapped into his face, half blinding him, but ahead he could still make out the white half-bald head, the thin hair washed flat across the dead face.

As soon as he reached him and started to pull him back, a rope landed in the water beside him. He grabbed it and together they were hauled towards the boat.

'All right, I've got him,' said Copper. The drowned man was heaved first into the boat where the line was secured, then up the ship's side where the helpers clung to the upper meshes of the net to take the weight on the way up.

McGill was feeling so strong that he scarcely needed Copper's assistance to get back into the boat. He got a knee into the looped lifeline that ran round below the gunwale, and landed spluttering on one of the thwarts.

Copper said, 'That's the stuff, lad!'

The other two men shouted, 'Come on now – that's the lot.' They all jumped for the net and climbed it as fast as they could. Below them, the boat pounded again with a splintering crash, its last frustrated effort to crush them. At the top they were engulfed in the crowd.

'All right – let the boat go – cut the grass as far down as you can.' The First Lieutenant was giving the orders. He turned to go up on to the bridge. 'Well done, Chapman – and you, Piggott. Who else was down there? Oh, well done, Money! You'd better go forrad and get your things off.'

The Bosun's Mate came up to him. 'From the Captain, sir. He thinks there's a man still on the ship. He says will you speak to him about it at once, sir.'

A man on the ship. A man trapped there in that burning hell. The group stood gazing, and McGill, dripping and shivering, stood with them.

As they watched, a torch flickered at the stern of the ship.

The First Lieutenant came back and spoke to the Petty Officer. 'The survivors say that there may be two men still on the poop there. The Captain wants us to float a Carley raft away on the grass line. We're drifting so fast that it will float up wind to them quite quickly ... No, no one's to go in it. We'll use No. 4 raft; better slip it at once.'

McGill heard all this; heard and understood it. But why was no one to go in it? With someone in it the trapped men could throw down a line from their quarter which could be caught on the raft. Otherwise it was pure chance if the raft went within their reach.

These were the sort of things that McGill knew – much more than he knew about sounding machines, carpenters' stoppers, and the breaking strains of cables, and all the other things you were meant to know in the Navy. The very fact that he knew and understood it was a challenge. Being cold and wet made the effort harder, until the challenge was irresistible. He was exultantly ready to do much more than his share.

The raft was slipped and fell with a big flat splash into the smoky blackness below.

McGill followed quickly to the guard-rail, climbed through and seized hold of the grass rope. Then he slid down into the sea and worked his way along to the raft at the far end of it. The water was desperately cold; much colder than before. He climbed eagerly on to the raft. Already it was drifting away from the destroyer's side; or rather the destroyer was being blown away from the raft, which floated low in the water, waves breaking over its rounded end. The centre of the oval raft formed a sort of basket, with its bottom in the water like a gigantic fisherman's keep net.

McGill sat on the cylinder with his feet in the basket, submerged to the knees. As he sat, he was suddenly seized with panic. Little feathers of spray burst over him. All at once he was tired and weak: his muscles went stiff, he was utterly alone in the dark ocean between the two ships.

The weakness and the fear held him in a tight grip so that his stomach ached sickeningly. Now he would surely die, and for such stupidity, such stupidity, such mad vanity! People could not live long like that – all wet and exposed. He would be washed off the raft soon. Exposure – people died of it, he had heard about it. They went numb. Someone had said it was a comfortable, easy death, once you were numb. What nonsense some people did talk.

'Don't get above yourself,' his mother used to say. That was not nonsense. But this that he was doing was nonsense – folly – madness. Only Tim would understand why he had started out.

Tim *would* understand. Like that night by Mount Stewart there, when they went on for the baits after the tide had turned just to show they weren't frightened. Well, now it was all over; he would surely die, surely.

The sea's ruffled blackness was velvet, the cloak of death.

'Is that a man in the raft?' called the First Lieutenant, as they payed out the grass line from the iron deck.

'Yes, sir,' answered the Petty Officer.

'Who is it?'

'I'm not sure, sir. I think it's McGill.'

Half a dozen voices confirmed it.

'I said no one was to go.'

'I never saw him go, sir.' And then, after a pause, 'He was the first man down into the boat, sir, and swam to get that drowned fellow afterwards. Seems a good lad, sir, in an emergency, as you might say.'

The grass line floated away to the black outline that was only occasionally visible when a wave crest lifted it against the glow of the flames beyond. The raft was certainly working towards the bows. With the fire amidships there was no chance of the trapped men reaching it there.

The First Lieutenant went up to the bridge.

'Everything going all right, No. 1?' asked the Captain.

'Well – yes, sir. Actually, sir, there's a man gone over in the raft.'

'I said no one was to go. Who is it?'

'Ordinary Seaman McGill, sir.'

'Yeoman, shine the ten-inch on the raft.'

In the beam of the searchlight, the little raft and the figure perched at one end looked very white shining out of the mist of spray. The figure was moving – swaying rhythmically.

'What the hell's he doing? Paddling?'

'Yes, sir.'

Paddles were kept lashed to the floor of each raft. With one of these McGill was trying as hard as he could to work himself towards the stern of the sinking ship, so that as it drifted down on him he would come up against it on the sternward side of the midship blaze.

The exercise was warming him and with the warmth his strength returned. The grass rope was not floating and acted as an anchor to the raft, but every inch was worth the effort. He felt much better and despised his fear. He was well in the lee of the burning ship, and the spray was not breaking over him so much, but the pungent burning smell filled his nostrils.

He was out of breath soon but he saw that he had gained a few yards and paddled on with a second wind. He was very near the other ship now – close above, the fire raged. He could feel the heat of it, like the heat of a summer sun; oh, the deadly sweet taste of exhaustion and the fear of it! Then, from aft, he heard a thin faint cry. At the same moment the coils of a rope splashed down near him.

He reached out with the paddle but it was too far and he

wondered whether he should jump in and fetch it. He decided to let them throw again, and at the second throw the line fell across the raft.

At once he made it fast and the raft was hauled towards the ship. The great stern fell with a thunderous slap and then rose high above him so that rudder and screw were clear of the water; the foam boiled around them as the ship pitched back. There was a Jacob's ladder hanging from the poop. If the raft was to reach the foot of the ladder, it would lie dangerously close to that plunging propeller, but there it would have to go. With the great waves the raft rose and fell past twenty or more rungs of the ladder.

One of the men had already started down, very slowly and hesitantly.

When he was a little more than halfway McGill shouted, 'Stay there now, and jump next time I come up to you.'

The man obeyed, and when McGill said, 'Now, jump!' he waited for about three seconds, and then jumped and missed. McGill reached out for him and pulled him into the middle of the raft; he sat on the edge, coughing.

It was darker now because they were shaded from the fire by the bulk of the stern. The searchlight had been doused, but even in the dark McGill saw at once that he had rescued a boy no older than himself. The boy was very frightened; he was sobbing with cold and fear, between fits of coughing.

This discovery made McGill feel suddenly old and competent and strong. With an effort he overcame the chatter of his teeth. 'It's all right,' he said, 'we'll be back safe directly.' He did not doubt that it was true, and he found that it encouraged him a lot to say it.

The second man was coming down now. As he came a huge wave lifted the ship's stern high out of the water, showing again the rudder and the propeller.

McGill thought, What if a bight of the grass line should catch under the propeller, or a loop go under the rudder? It would pull the raft down. He thought, It will do, it is bound to; and then it will really be the end.

He clambered along the raft and saw how the grass rope ran steeply down now into the water. It had sunk deep and the ship was drifting down on it. He tried to haul some of it in. 'Hey, give me a hand!'

But the boy did not move, he held on with both hands, too frightened to help. McGill had to give up. If the raft was pulled down they would drown. There was nothing he could do. He was quite resigned.

The second man stepped off neatly into the raft. He was frightened too.

'Are there any more?' McGill shouted.

'No,' he heard the man say. 'No, we're the last.'

With his knife he cut the rope which hung from the bulwarks far above. All that now remained was for the destroyer's men to haul them back the few hundred yards which separated the two ships. That was all, if the rope was not foul of the derelict. But how could they know in the destroyer that the raft was ready? McGill shouted, 'Heave in,' but the sound was drowned instantly in the fury of the night.

Then he remembered the flickering torch.

'Have you still got your torch?' he asked.

The man fumbled in his pocket and produced it. 'It's pretty well finished,' he said.

McGill took it and flashed once or twice. Then pressing on the button without pushing it forward he started to signal, out of the far-off memories of the training establishment. H – that was four dots; E – oh, E was easy, just one dot; and then A – a dot and a dash; V – that was difficult, he paused and then made a B – a dash and three dots instead of three dots and a dash; then E again. At the end of the word he had to wait to see if it had been read.

No answering flash came from the destroyer's bridge.

The torch was almost spent, showing only a dull amber light.

He made the word all over again, but he was sure they could not see it.

And then suddenly a point of light showed on the bridge. T, it said, which meant that it had understood. IN was easy to make: I – two dots, and N the opposite of A – a dash and a dot. The light answered again, paused, then fluttered quickly, far too quickly for him to read, but he guessed it meant that they had understood.

The man and the boy sat very still, clutching the rope strops of the raft. The boy was shivering; his breath came in quick sobs.

McGill climbed along to the other end of the raft and watched the grass rope which hung vertically under it. Suddenly, he looked up and noticed they were away from the ship's side, and at the same time he determined that the grass rope was not quite so vertical as it had been; it sloped away towards the destroyer; it was *not* round the screw of the wreck. They were being hauled back.

'Won't be long now,' he said quite cheerfully.

The man looked at McGill, for the first time aware that he was only a boy. He saw his black hair tangled across his forehead, and the glint of the fire shining in his eyes; he saw his shoulders hunched and shivering, and the thinness of him. 'Good,' he answered through clenched teeth.

McGill sat quite still, gazing back at the burning wreck.

'Look!' he said suddenly.

Her back was broken and her two black ends were rising up. While they watched there came an awful hissing roar as the water boiled up and quenched the red-hot plates. There was a muffled explosion and a great shower of sparks burst out of the centre of the fire. A pall of billowy smoke shone blood-red for a moment, and then faded as the blaze was drowned. The stern and bows rose majestically into the air, almost touching as the ship folded in half. Then, quite silently, she slid down behind the black mountain of an intervening wave.

When next the raft rose upon a crest, nothing remained but

a flicker of flame burning on the sea's surface. It flared for a moment and went out.

McGill looked back to the destroyer, a dim shape in the new darkness. He was utterly unmoved by the final plunge of the sinking ship. Too much had happened already. He was past caring about anything, numbed with awful cold and fatigue.

They came halfway down the net to pull the three of them from the raft. They had to hold McGill up and help him along to the mess deck. Copper was on one side of him. 'Ee – that's the stuff, lad,' he said.

They helped him off with his sodden clothes and rubbed him down with towels, and someone brought him a steaming cup of cocoa with too much sugar in it. The First Lieutenant was there some of the time, saying something about it being a good show.

McGill was too tired to remember it all. They wrapped him in blankets, and bundled him into his hammock. A rich glow came into his body and he curled his hands up to his face.

'Oh, Tim,' he said to his brother, 'it was all right this time, wasn't it?'

A moment later, he was asleep.

From *Argosy*

30

Camouflage

Shortly afterwards we camouflaged the *Broke* on my scheme. It was designed to make her as pale as possible at night because ships nearly always appear as dark blobs on the horizon against the sky. Only down moon on a very bright night would she look too pale. Otherwise she could hardly be white enough. Accordingly one side was a compromise dazzle scheme. The best dazzle schemes, for deception, deceiving the enemy about the class of ship, the speed and angle of approach or inclination, are those which have contrasting patches indicating edges and surfaces which do not exist. The best contrast is one of tone – that is to say, black and white. But black or very dark tones ruin the night invisibility idea. At night, of course, colour is not visible, therefore provided that the colours are pale and of the same tone, the ship will look pale grey all over no matter what colours are used. At the same time, within the scope of paleness, bright colours juxtaposed may, in bright sunlight, produce the desired dazzle effect suggesting nonexistent edges and false planes. The other side of *Broke* was therefore painted in pale pastel shades of blue, green, buffish pink and white.

The early reports from our sub-divisional mate *Vansittart*'s Captain, Lt-Commander Evershed, who years before had once written me a fan letter about my book, *Wild Chorus*, were most encouraging. At dawn and dusk too, he reported that the ship was remarkably invisible.

—

The whites and pales greys and blues in which I had first painted *Broke*, following my conviction that night invisibility was of value in the U-boat war, had now spread to many other ships. In the first instance *Broke* had been compared with destroyers painted a curious purplish mauve colour. These destroyers belonged to the Fifth Destroyer Flotilla commanded by Captain Lord Louis Mountbatten and the colour, having been his idea, was called 'Mountbatten pink'! It was based, I believe, on the invisibility of a Union Castle liner on a particular occasion. At night it was better than the standard grey because it was a little lighter, but the pink colour (as opposed to the tone) did not seem to be of any significance. I stuck to my original opinion that there was nothing magic about Mountbatten pink except the name.

Meanwhile the then C.-in-C. Western Approaches – Admiral Nasmith – had been sufficiently convinced by my arguments in favour of the pale blues, pale greys and whites, to commission me to design camouflage schemes for the fifty obsolete American destroyers, which we had acquired in exchange in the fall of 1940, long before the United States entered the war. Soon, however, all the destroyers in the Western Approaches Command had turned near-white. It must have been a little before this that a question had been asked in Parliament: Was the First Lord of the Admiralty aware that ships painted white had been allowed to sail in convoy thereby endangering all the other ships in the convoy? . . .

But the idea was now becoming generally accepted. At night no paint could ever make a ship *whiter* than the sky behind it. A white ship would match it most nearly, and be less easily seen than a dark one.

I wrote the following notes on the scheme at the request of C.-in-C. Western Approaches:

CAMOUFLAGE OF VESSELS OPERATING AGAINST U-BOATS

1. The first all-important principle is its operational object. In this case it is to avoid being seen by a U-boat on the surface at night. The scheme must be designed for this purpose, and this purpose alone; all questions of rendering a ship less liable to successful torpedo attack by confusing dazzle painting; and realistic false bow waves, etc., are considered, at the present stage of U-boat tactics, to be irrelevant.
2. Compromise is usually fatal in a camouflage scheme. Invisibility at night must be the *only* objective.
3. To make a ship less visible at night it must be painted a very pale tone, especially that part of it which is likely to be seen from conning-tower level against the sky.
4. Any one scheme cannot be completely successful under all weather conditions, therefore the best plan must be to seek invisibility in conditions which obtain on the majority of nights in the north-west approaches.
5. The very pale ship is not effective when lit directly by a bright moon, but it is effective on all dark nights and on all overcast moonlight nights; on bright moonlight nights the pale ship will still be better than the dark one against the up-moon half of the sky, only worse in the down-moon half, a fifty-fifty chance. Fifty per cent of the dark hours are moonlit, but a large number of the moonlit hours are overcast. Every overcast moonlit hour weighs the scale in favour of the pale camouflage scheme, as the most suitable for the majority of the conditions.
6. A number of variations of the pale scheme have been tried, and have proved successful under the conditions described above. It has turned out by chance that they are also effective under certain day conditions and at dawn and dusk.

7. When, under bright direct moonlight the pale scheme makes the ship show white, it still has a slight advantage over the ship that shows black, as nearly all ships do. This advantage is the psychological one that all look-outs are expecting to see a ship as a dark mass. Recently a look-out reported to me a 'Town'-class destroyer (painted with a form of the pale scheme) under a bright moon, about five cables away as 'something making white smoke on the surface'.
8. In conclusion the scheme should work to the best advantage on occasions when a U-boat might be surprised on the surface, and to make the *Vanoc*'s feat [the destroyer *Vanoc* had recently sunk a U-boat by ramming] more likely to be repeated, or on occasions when a U-boat might select what he thought was a gap in the screen, and find to his mortification that there was in fact an escort that he had not seen, right ahead of him.
9. It is emphasised that the variations within the range of the 'pale scheme' are still experimental and capable of improvement.
10. Notes on the general application of the scheme which have been issued for the guidance of First Lieutenants of ships I have camouflaged are appended.

P. Scott, Lieutenant, R.N.V.R.

By early 1941 C.-in-C. Western Approaches was making the following signal to Captains (D) Londonderry, Greenock and Liverpool, repeated Admiralty, Flag Officer Commanding North Atlantic, C.-in-C. South Atlantic:

> Four corvettes having been torpedoed up to date every endeavour is to be made to hasten camouflaging of corvettes in accordance with Peter Scott scheme.

But then a new factor arose. If the enemy could not see the escorts there was an improved chance of surprising U-boats, but if the escorts could not see each other there was also an increased chance that they would collide. *Broke* and *Verity* were not the only escort ships to be involved in collisions for which good camouflage was accepted as a primary cause. The Admiralty called a conference to consider whether these disadvantages outweighed the advantages, and finally came down on the side of the camouflage. I made designs for escort trawlers, and later for the Motor Gunboats and Motor Torpedo Boats of Coastal Forces. I wrote memoranda on the theory and practice of ship camouflage for Combined Operations. I wrote practical hints for First Lieutenants who had to paint their ships. With these First Lieutenants I was far from popular, for my light colours showed dirt and rust immediately, whereas against Mountbatten pink they were hardly noticeable.

From *The Eye of the Wind*

31

Fire at Sea

On Sunday 6 April 1941, I came on watch at 4.00 p.m. for the first dog watch and found a wild grey day of heavy seas, driving squalls, and a moderate south-easterly gale.

We [HMS *Broke*] were on our way to meet a homeward-bound Gibraltar convoy, having parted company with the outward-bound convoy in the early morning, and we were plugging into a head sea at seven knots which was as much as we could do with comfort.

We were in company with *Douglas* (an old Destroyer Leader like ourselves, and Senior Officer on this occasion) and *Salisbury* (an ex-American destroyer) and our position was on the port wing of a sweeping formation.

When I came on to the bridge three other ships were in sight, which was unusual because chance meetings were not common in 21° West – six hundred miles out in the Atlantic – more especially on days of poor visibility like this one.

These three ships were already some way astern and steering west, and I learned that we had 'spoken' to them half an hour before when they had first come into sight. They were HMS *Comorin*, ex-P. & O. liner of fifteen thousand tons, now an Armed Merchant Cruiser, the *Glenartney*, a smaller merchant ship, and *Lincoln*, another ex-American destroyer, who was their escort.

They soon disappeared into the smoky haze and my watch

passed uneventfully as I kept station on *Douglas* and contemplated, from the cosy depths of my duffle coat, the white foam streaks which patterned the grey mountainsides of wave, and the green curling crests that were the only traces of colour in that inhospitable scene.

About twenty minutes after I had been relieved I was informed in the wardroom that the starboard fore-topmast-backstay had parted, and I went at once to the bridge to organise the repair. While I was there a signal was received from *Lincoln* at 6.20 p.m.:

> HMS *Comorin* seriously on fire in position 54° 39′ N., 21° 13′ W. join us if possible. 1600/6.

We passed this to *Douglas* for permission to go and were detached at 6.40.

Having told *Lincoln* by radio that we were coming, we made a further signal: 'Expect to sight you at 2015. Is a U-boat involved?' to which came: 'Reply No.'

During this time we were able to increase to eighteen knots and with the following sea the ship's motion was so much easier that the backstay was quickly repaired and secured. We then began making preparations for what might turn out to be our sixth load of survivors since war began.

Scrambling nets, lifebuoys, oil drums (to pour oil on to the troubled waters), heaving-lines and heavier bow and stern lines for boats, line-throwing-gun, grass line and manilla – all these were got ready and the mess decks were prepared for casualties.

At eight minutes past eight in the evening *Glenartney* was sighted fine on the port bow at extreme visibility range – about eight miles – and four minutes later *Comorin* was seen right ahead, which was exactly three minutes earlier than expected. There was a lot of white smoke coming from her single funnel.

We asked *Lincoln*, who was lying just to windward of the burning ship, what the situation was and she told us that

Comorin was being abandoned, that no boats were left, that the *Glenartney* was picking up some rafts and that she, *Lincoln*, was hauling over rafts on a grass line.

When we drew near the scene was awe-inspiring. The great liner lay beam on to the seas drifting very rapidly. A red glow showed in the smoke which belched from her funnel and below that amidships the fire had a strong hold. Clouds of smoke streamed away from her lee side. The crew were assembled aft and we were in communication by lamp and later by semaphore. From the weather quarter the *Lincoln*'s Carley rafts were being loaded up – a dozen men at a time and hauled across to the destroyer lying about two cables away. It was a desperately slow affair and we went in close to see if we could not go alongside.

Various objects were falling from time to time from the *Comorin* into the sea. Each time one looked carefully to see if it was a man but it was not. Some of the things may have been oil drums, in an attempt to make the sea's surface calmer. We looked at the weather side first and passed between *Lincoln* and *Comorin*, only discovering that they were connected by line after we were through. But the line did not part, it had sunk deep enough for us to cross. We turned and tried to lay some oil between the two ships. This meant that we nearly got sandwiched. There were two raft loads going across at the time and the sailors in them waved us to go back. They seemed incredibly tiny and ant-like in their rafts on those angry slopes of grey sea.

The *Lincoln* was rolling wildly and once when her propellers came clear of the water we could see that she had a rope round the starboard one. Her starboard side was black, which we thought at first was burnt paintwork, but discovered afterwards was oil which they had tried to pump out on the weather side and which had blown back all over their side and upper deck.

To go alongside *Comorin* seemed an impossibility. The waves were fifty to sixty feet from trough to crest and the liner's cruiser stern lifted high out of the water at one moment showing rudder and screws and crashed downward in a cloud of spray the next. I thought a destroyer could not possibly survive such an impact.

Round on the leeward side we got our grass line out on to the fo'c'sle and dropped a Carley raft into the water attached to it. The liner drifted to leeward and so did we, but the raft with less wind surface did not drift so fast. So it appeared to float away up wind towards the other ship, and in a very short time was against the liner's lee side. There was a Jacob's ladder over the side and one rating went down and eventually got into the raft. We started to haul in on the grass but we had allowed too much grass to go out, the bight of it had sunk, the great stern had lifted on a wave and come down on the near side of the rope; the bight was round the rudder. When we hauled on the grass it pulled the raft towards the liner and under her. Each time the ship lifted it pounded down on the man in the raft, who began to cry out. At this time we drifted round the stern and could not see him any more. I am not certain if he was drowned, but it can only have been a miracle that saved him.

We were close to the stern now and we fired our Coston line-throwing gun. We fired it well to windward, but a little too high and the wind blew it horizontally clear of the ship like straw. But we were close enough to get a heaving-line across – after many had missed – and we connected the Coston line to the departing heaving-line as we drew astern and finally pulled an end of the grass rope to it and passed it over to the other ship. Then we put another Carley raft in the water so that they could haul it in to windward but for some time they made no effort to do this and finally their Captain made a signal that he thought the only chance was for *Broke* to go alongside and let the men jump. This was at 9.50 p.m.

I had various discussions with my Captain as to which side it should be. I must confess that I did not believe we could survive such a venture. By this time it was almost dark and the *Lincoln*'s raft ferry had failed owing to the parting of their grass line. I do not know how many people they had rescued by this ferry, but it cannot have been very many as it was desperately slow. Not that it wasn't enormously worth doing, for at the time it seemed to be the only way at all.

I saw one body floating away and wondered if it was the chap from our raft. Indeed I felt sure it was, though I heard a rumour afterwards that he had managed to climb back on board, and I know that several men were drowned embarking in the *Lincoln*'s ferry service. Obviously a raft secured to the upper deck of a heavily rolling ship cannot be secured close alongside. If it were it would be hauled out of the water by its end whenever the ship rolled away and all those already embarked would be pitched out. The rafts therefore had to remain ten or fifteen yards from the *Comorin*'s side and men had to go down the rope to them. It was here that several men were drowned.

As it gradually got dark the glow from the fire shone redly and eventually became the chief source of light. As soon as we knew we were going alongside I went down to the fo'c'sle, got all the fenders over to the port side, and had all the locker cushions brought up from the mess decks.

I suggested that hammocks should be brought up, but Angus Letty, our Navigating Officer, said that since it was doubtful if we should get alongside he thought it would be a pity to get them all wet. I wish I had pressed the point but I didn't. Letty had been doing excellent work on the fo'c'sle all the time, but I think his judgement was at fault in this and so was mine in not seeing at once that he was wrong. What are a few wet hammocks by comparison with broken limbs?

So we closed the starboard (leeward) quarter of the *Comorin* and in a few minutes we had scraped alongside.

The absolutely bewildering thing was the relative speed that the ships passed each other in a vertical direction. The men waiting on the after promenade deck were forty feet above our fo'c'sle at one moment and at the next they were ten feet below. As they passed our fo'c'sle they had to jump as if jumping from an express lift.

The first chance was easy. About nine jumped and landed mostly on the fo'c'sle, some on B gun deck. They were all safe and uninjured. As they came doubling aft I asked them how they got on and they were cheerful enough.

One petty officer came with this lot – P.O. Fitzgerald – and I immediately detailed him to help with the organisation of survivors as they embarked.

There was little, if any, damage to the ship from this first encounter and we backed away to get into position for the next. As we closed in for the second attempt the terrific speed of the rise and fall of the other ship in relation to our own fo'c'sle was again the main difficulty for the jumpers, added to which the ships were rolling in opposite directions, so that at one minute they touched, at the next they were ten yards apart. This very heavy rolling made it almost impossible for us to keep our feet on the fo'c'sle. We had to hold on tightly nearly all the time. I remember I had started the operation wearing a cap as I thought I should be easier to recognise that way, but I so nearly lost it in the gale that I left it in the wireless office for safe keeping.

The second jump was much more difficult than the first. Only about six men came and three of them were injured. From then on I do not remember the chronology of events as one jump followed another with varying pauses for manoeuvring. Our policy was not to remain alongside for any length of time as this would have been very dangerous and might well have damaged us to the point of foundering in these monstrous seas. Instead we quickly withdrew after each brief contact in order to assess the damage and decide upon the seaworthiness of the ship before closing in again for another attempt.

The scene was lit chiefly by the fire, as it was now pitch dark. Occasionally the *Lincoln*'s searchlight swept across us as they searched the sea for the last Carley raft which had somehow broken adrift. At each successive jump a few more men were injured.

We decided to rig floodlights to light the fo'c'sle so that the men could judge the height of the jump. These were held by Leading Telegraphist Davies and Ordinary Seaman Timperon all the time. The pool of light which they formed gave the whole scene an extraordinary artificiality, as if this were some ghastly film scene, and I can remember, as I stood impotently waiting for the next jump, feeling suddenly remote as if watching through the wrong end of a telescope.

As soon as the jumping men had begun to injure themselves I sent aft for the doctor who was in bed with a temperature. He came for'ard at once and started to work. At the third jump we drew ahead too far and the whaler at its davits was crushed but several men jumped into it and were saved that way. Another time the Captain changed his mind and decided to get clear by going ahead instead of astern. The ship was struck a heavy blow aft on the port for'ard depth charge thrower.

At this jump there had been a good many injured and as we went ahead into the sea again to circle *Comorin* for another approach the decks began to wash down with great seas. This was awkward for the disposal of the injured and I went up and asked the Captain if in future we could have a little longer to clear away and prepare for the next jump – to which he readily agreed.

My routine was to be down by A gun for the jump. There was a strut to the gun-cover frame which was the best hand-hold. Here I saw that the padding was properly distributed, the light properly held, the stretcher parties in readiness and hands all ready to receive the survivors.

Then we would crash alongside for a few breathtaking seconds whilst the opportunity did or did not arise for a few to jump. Sometimes there would be two opportunities – first while we still went ahead, and then again as we came astern. The great thing was to get the injured away before the next people jumped on to them. As soon as this had been done I went with the Shipwright to survey the damage and then up to report to the Captain. This was a rather exhausting round trip and before the end I got cramp in my right arm through over-exercise of the muscles.

The damage to the ship was at first superficial, consisting of dents in the fo'c'sle flare and bent guard-rails and splinter shields. But at length one bad blow struck us near the now-crushed whaler on the upper deck level. I ran aft and found a large rent in the ship's side *out* of which water was pouring. I did not think the boiler room could have filled quickly enough for this to be sea water coming out again, and came to the conclusion that it was a fresh-water tank. This was later confirmed by the Shipwright. As I had thought, it was the port 'peace' tank. This was an incredible stroke of good fortune, as a hole of that size in the boiler room would have been very serious indeed in that sea.

Several of *Comorin*'s officers had now arrived and we began to get estimates of the numbers still to come. There seemed always to be an awful lot more. I detailed Lt Loftus to go down to the mess decks as chief receptionist and went back to my usual round – fo'c'sle for a jump, clear the injured, view the damage, report to the Captain and back to the fo'c'sle.

By now some of the injuries appeared to be pretty bad. There were a good many broken legs and arms and one chap fell across the guard-rail from about twenty-five feet. Letty came aft to me and said 'That fellow's finished – cut his guts to bits.' It appears that A.B. George, a young Gibraltar seaman who was doing excellent work on the fo'c'sle, had put a hand on his back – felt what he took to be broken ribs and withdrawn his hand covered with blood. No such case ever reached the doctor and although there is a possibility that an injured man could have gone over the side owing to the heavy rolling, I am inclined to think that the account was an exaggerated one.

It filled me with gloom, and since at least one-third of the survivors landing on the fo'c'sle had to be carried off it, I became desperately worried by the high percentage of casualties. True, they had improved lately since we had ranged all available hammocks in rows, like sausages. They made very soft padding but a few ankles still slipped between them and got twisted.

On one of my round trips I met an RNVR Sub-Lieutenant survivor trying to get some photos of the burning ship (I never heard whether they came out). There were a good many officers on board by this time, mostly RNR. Of course they did not recognise me, hatless in duffle coat, grey flannel trousers, and sea boots, and it took a few moments to persuade them to answer my searching questions.

Still the estimated number remaining did not seem to dwindle. Sometimes there were longish pauses while we manoeuvred into position. Twice we went alongside without getting any men at all. Once the ship came in head on and the stem was stove in, for about eight feet down from the bull ring, and we got no survivors that time either.

Somewhere around 11.30 p.m. the fire reached the rockets on the *Comorin*, which went off splendidly, together with various other fireworks such as red Verey's Lights and so on. Later the small arms ammunition began to explode, at first in ones and twos and then in sharp rattles and finally in a continuous crackling roar.

The sea was as bad as ever and at each withdrawal we had to go full astern into it. This meant that very heavy waves were sweeping the quarter-deck, often as high as the blast shield of X gun (about eight feet). When these waves broke over the ship she shuddered and set up a vibration which carried on often for twenty or thirty seconds. All this time the hull was clearly under great strain. As we got more used to going alongside some of my particular fears grew less, but the apparent inevitability of casualties was a constant source of worry and there was also the continual speculation upon how much the ship would stand up to.

Once on the first approach there seemed to be a chance for a jump. Two men jumped, but it was too far and they missed. I was at the break of the fo'c'sle at the time, and looked down into the steaming, boiling abyss. With the two ships grinding together as they were there did not seem the slightest chance of rescuing them. More men were jumping now and theirs was the prior claim. The ship came ahead, men jumped on to the flag deck and the pom-pom deck amidships was demolished: then as we went astern, I ran again to the side to see if there was any sign of the two in the water, but there was none that I could see.

Having got the injured off the fo'c'sle I went aft to examine the damage to the pom-pom deck and see if the upper deck had been pierced. I heard a very faint cry of 'Help', and looking over the side saw that a man was holding on to the scrambling nets which I had ordered to be lowered as soon as we arrived on the scene. We were going astern but he was holding on. I called to some hands by the torpedo tubes and began to haul him up. Eventually he came over the guard-rail unconscious but still holding on by his hands; his feet had never found the net at all. He was very full of water but we got him for'ard at once and he seemed likely to recover.

About this time too they had an injured man on X gun deck who was got down by Neill Robertson stretcher (the kind in which a man can be strapped for lowering). These stretchers were being used all the time. I went for'ard once to find a petty officer. There was a big crowd outside the Sick Bay and I trod on something in the dark shadow there. It was an injured man and I had tripped over his broken leg. I hope and believe it was still numb enough for him not to be hurt, but it was very distressing to me to feel that I had been so careless. I ought to have known that the congestion of stretchers would be there.

When I was reporting damage to the Captain on one occasion we were just coming alongside so I stayed on the bridge to watch from there, as there was hardly time to get down to the fo'c'sle. The Captain was completely calm. He brought the ship alongside in the same masterly manner as he had already done so often. He was calling the telegraph orders to the Navigating Officer who was passing them to the Coxswain in the wheelhouse. As we ground alongside several jumped. One officer was too late, and grabbed the bottom guard-rail and hung outside the flare, his head and arms only visible to us. There was a great shout from the crowd on *Comorin*'s stern. But we had seen it and already two of our men (Cooke and George) had run forward and were trying to haul the officer on board. The ships rolled and swung together. They would hit exactly where the man dangled over the flare; still the two struggling at the guard-rail could not haul the hanging man to safety. Then as if by magic and with a foot to spare the ships began to roll apart again. But still the man could not be hauled on board; still he hung like a living fender. Again the ships rolled together and again stopped a few inches before he was crushed. As they rolled towards each other for the third time, Cooke and George managed to get a proper hold of the man and he was heaved to safety and this time the ships crashed together with a rending of metal. The two seamen had never withdrawn even when the impact seemed certain and they had thus saved the officer's life. It was a magnificently brave thing to see.

Another man was not so lucky. In some way at the time of jumping he was crushed between the two ships, and fell at once into the sea as they separated. On the other hand, I saw a steward with a cigarette in his mouth and a raincoat over one arm step from one ship to the other, swinging his leg over the guard-rails in a most unhurried manner, just as if he had been stepping across in harbour. It was one of those rare occasions when the rise and fall of the two ships coincided. One man arrived on board riding astride the barrel of B gun and another landed astride the guard-rail. Both were quite unhurt.

During all this long night Lewis (our Gunner 'T') was with me on the fo'c'sle. He was a great stand-by, always cheerful and helpful. As we saw the damage getting gradually worse we found a chance to speculate on the leave we should get and even to hope that the ship would not be 'paid off' because of the terrible paper work involved and the number of things that each of us had 'on charge' in our respective departments which we knew would never be found and for which we should have to account. Letty was fully of activity too on the fo'c'sle head and was doing excellent work among the survivors as they arrived on board, tending the injured, heartening the frightened, organising the stretcher parties.

After the terrible vibrations set up by the huge waves which broke over the quarter-deck each time that we had to go 'Full astern both', I became anxious about the state of the after compartments and sent the Gunner and the Shipwright aft to find out if there was any extensive flooding. They had not long been gone when one of our officers came up and said that there was six feet of water in the wardroom. I was inclined to take this with a grain of salt and it was lucky that I did so, for when the Shipwright returned he reported that there were a few inches of water slopping about on the deck but nothing whatever to worry about in any of the after compartments. The officer's explanation was that someone had told him this, he didn't know who, and he told this 'someone' to go aft and verify it and then report to me. Nothing of course had been reported to me.

This same officer only impinged himself once more upon my consciousness during that night. That was when he came up and told me in a rather panicky tone that the damage was really getting very bad on the mess decks and that he did not think that the men would stand much more of it. His explanation of this remarkable contribution was that an Engineer Lt-Commander survivor who was down there and evidently (and, I hasten to add, with every justification) slightly shaken, had told him that he had better report the extent of the damage to his First Lieutenant.

Actually, of course, all this damage was above the water line, although, owing to the rolling and pitching a good deal of water was coming inboard through the numerous rents along the port fo'c'sle flare.

Back on the upper deck we saw a man working at the edge of the fire on the weather side of the *Comorin*'s main deck. We couldn't make out what he was doing, but we discovered afterwards that he was burning the confidential books in the blazing signalmen's mess.

Still we were periodically closing the *Comorin*'s bulging stern, with the wicked-looking rudder and propellers bared from time to time as the massive black shape towered above us. Still the men were jumping as the opportunity arose. Each time there was a chorus from the fo'c'sle of 'Jump! – Jump!'

Although we were keen to persuade as many to jump as possible, I thought that it was better to let them judge the time for themselves and told my men not to call out 'jump'.

From one of the survivor officers I discovered there had been two medical officers in *Comorin* – one the ship's doctor, Lt-Commander, RNVR, and another taking passage, a Surgeon Lieutenant, RNVR, but the latter was believed to have gone in one of the boats.

So next time we went in to the *Comorin* I shouted that I wanted the doctor at the next jump if possible, as I knew that our own doctor would be hard pressed with so many casualties. No chance arose that time, but a little later as we closed again I could see the Surgeon Lt-Commander poised on the teak rail waiting to jump. Here was the one man that I really wanted safe but he missed his best opportunity and took the last chance, a jump of twenty feet. He landed beyond the padding, at my feet as I stood on the layer's platform of A gun. He fell flat on his back and lay there quite still. But in a few moments he had come round and ten minutes later, very bruised and stiff, he was at work amongst the injured. He too had barely recovered from 'flu and was quite unused to destroyers; I believe also that he felt very seasick. His work that night was beyond praise.

Casualties were not so heavy now, but there were still a good many owing to the fact that jumpers on the downward plunge of the *Comorin* often met the upward surge of *Broke*'s fo'c'sle.

This meant that they were hurled against the deck like a cricket ball against a bat.

We were all soaked through with rain and spray and sweat. It was very hot work in a duffle coat. My arms were still inclined to get cramp if I went up ladders by pulling instead of climbing. However I don't think this interfered with my work at all. I still continued with the wearisome anxious round – fo'c'sle for the jump, assess damage, up to report it, and back to the fo'c'sle for the next jump.

These last jumps were very awkward because the remaining ratings were the less adventurous ones whom it was difficult to persuade to jump. Two of them, senior petty officers, were drunk. They had been pushed over and arrived on board in an incapable condition. I thought they were casualties but was relieved, and at the same time furious, to discover that they were only incapably drunk. It seemed too much that my stretcher parties should work on two drunken and extremely heavy petty officers, especially as they were the very two on whom the officers in the other ship would be expecting to be able to rely to set a good example to the rest.

Once I sent up to the Captain and suggested that we might perhaps consider rescuing the last few by raft if the damage got any worse, as with well over a hundred survivors on board it would be out of proportion to risk losing the ship and all of them to get the last few if there were a good chance of getting those last few in another way. We agreed that we would try this as soon as the damage gave real cause for alarm but not before.

Quite suddenly the number on *Comorin*'s stern seemed to have dwindled. At last we seemed to be in sight of the end. There were about ten, mostly officers. The Captain and Commander were directing from one deck above, with a torch.

At the next jump they all arrived, unhurt, except the Captain and the Commander, who remained to make sure that the ship was clear.

Five minutes later we came in again. The Commander

jumped and landed safely. The Captain paused to make sure that he had gone, then jumped too. He caught in a rope which dangled from the deck above, and which turned him round so that he faced his own ship again. At the same moment the *Broke* dropped away and began to roll outwards. The Captain's feet fell outside the guard-rail and it seemed that he must be deflected overboard. But he sat across the wire guard-rail and balanced for a moment before rolling backwards and turning a back somersault on the padding of hammocks. He was quite unhurt, smoking his cigarette, and he had managed to return his monocle so quickly to his eye that I thought he had jumped wearing it.

He turned out to be Captain Hallett – a destroyer Captain of the First War who had served under my father some thirty-five years earlier.

I took him up to the bridge to see my Captain and at forty minutes past midnight we made the signal to *Lincoln*, 'Ship is now clear of all officers and men.'

Rescue-operations were now successfully completed. During the past three hours no less than 685 telegraph orders had been passed from the bridge to the engine room and executed without a mistake. Captain Hallett told my Captain that there were still some confidential books in the strong room of the *Comorin* and that he would not feel safe in leaving her until she had sunk. He asked us to torpedo her. Having discussed the possibilities of salvage and decided that they were not practical, we prepared to fire a torpedo. It was fired at rather too great a range and did not hit. Nor did the second or the third. Some time elapsed between these attempts and I went below to examine the situation on the mess decks. 'Anyone here down-hearted?' I asked and received a rousing cheer of 'No.' A lot of water was slopping in through the numerous holes and I started a baling party with buckets. The casualties were mostly in hammocks, some on the lockers and some in the starboard hammock netting. The port side of the mess deck was a shambles. However, since everything seemed to be under control, I went aft to the wardroom and found some thirty-five survivor officers ensconced there. The deck was pretty wet and the red shellac colouring had started to come off so that everything soon became stained with red. The flat outside was running with fuel oil and water and baling operations were put in hand.

Then I went back to the bridge and instituted a count of the numbers of survivors which turned out to be correct and never had to be altered. We had exactly 180 on board. Back in the wardroom I found the Engineer Officer baling away with buckets. He told me that the Gunner had been washed overboard, but had been washed back on board again by the next wave. He was very shaken and had turned in. I discovered also that earlier in the evening when struggling with the manilla, P.O. Storrs had been pushed over by the rope, but had managed to hold on and be hauled back to safety. At some time or other A.B. Bates had been washed off the iron deck just by the for'ard funnel and washed back on board again on the quarter-deck, a remarkable escape.

As it was now after 2.00 a.m. and I had to go on watch at 4.00 I tried to get a little sleep in Jeayes's cabin. My own cabin was occupied by a Commissioned Gunner whose arm had been crushed. He had caught hold of the outside of the flag deck. His arm had got between the two ships and been pinched whereupon he had let go and would have fallen into the sea but for the fact that his raincoat was pinched in and held. He hung by it and was hauled in from the flag deck ladder. I afterwards discovered that he had served with my father as an A.B. in the *Majestic* in 1906 when my father had been Cable Officer in her. He occupied my cabin for the rest of the voyage.

I did not sleep much and I missed the torpedoing of the *Comorin*, which was achieved with either the fourth or fifth torpedo. The sixth had a seized-up stop valve and could not be fired. At 4.00 I relieved Jeayes on the bridge. The *Comorin* had been sunk; she was appreciably lower in the water and had a marked list to port instead of the slight one to starboard which she had had during the rescue. The Captain had turned in in the charthouse, but came up for a few minutes. I was on watch by myself and we were patrolling up and down to weather of the blazing wreck until dawn.

By about 5.30 the whole of the after promenade deck of *Comorin* from which the survivors had jumped was ablaze. It was satisfactory to know that we had been justified in attempting the night rescue and that time had been an important factor. At dawn we left *Lincoln* to stand by *Comorin* till she sank and set off home with *Glenartney*.

The return journey entailed a lot of hard work. Baling was continuous for the whole two and a half days. *Glenartney* was ordered to proceed on her journey calling at Freetown to disembark survivors, and we were left alone. The weather worsened and on Monday night we had to reduce speed during the middle watch from eight to six knots owing to the heavy head sea set up by the easterly gale. The water on the mess decks was more than the balers could cope with and was running out over the watertight door sills in waterfalls. There were about sixty tons of water on the upper mess deck – about one foot six inches of water when the ship was on an even keel, and four feet or more when she rolled, with a corresponding cataract when she rolled the other way. Life on the mess decks became very uncomfortable. The forepeak and cable locker were flooded and the fore store was filling. It was an anxious night. But when I came on watch at 4.00 a.m. on Tuesday the gale began to moderate suddenly. One of the survivor officers – an RNR lieutenant – was keeping watch with me. By 5.00 the wind had died away to nothing and an hour later it began to blow from the west. Almost at once the sea began to go down and we were able to increase speed. By noon we were steaming at twenty knots for the Clyde and we arrived there on the morning of Wednesday 9 April. I spent those two and a half days mainly amongst the survivors for'ard. Lewis and the others were looking after the officers as well as possible in the wardroom. The for'ard mess deck seemed to need most of my efforts, and besides I enjoyed the company of the rating survivors, so I spent as much time as I could on the mess deck amongst them. The *Comorin*'s Commander co-operated most kindly by providing three watches of baling parties with two officers in charge of each watch and by helping with other work. Three survivor officers shared the night watches with us on the bridge.

In spite of the apparently heavy casualties during the rescue, so many of these recovered quickly that there were finally only about twenty-five hospital cases out of the whole 180. The worst cases were compound fractures. There were three legs and one arm. The worst of these cases was an old Chief Stoker who was sixty-nine years old. He was in bad pain all the time and to begin with he had been placed on the lockers in the doorway on the starboard side of the mess deck, from which he had to be moved later on. He was remarkably brave. The doctors were afraid he would die of shock, but on the last morning I had a half-hour's talk with him in which he described in spirited terms how he had once been presented to Queen Victoria. I believe he survived.

I found the youth whom we had pulled up on the net; he was lying stark naked in a hammock. He was still a bit under the weather but I got his clothes brought up from the boiler room and made him sit outside in the afternoon. Next day he was quite well and very bright. He was only eighteen and had been one of Carroll Levis's discoveries. His name was Sturgess and when he left he said, 'Thank you, sir, for saving my life – I'll put in a request as soon as I get to depot to join your ship.'

We arrived at Greenock at about 10.30 in the morning and were alongside for rather less than one hour before slipping and proceeding to Londonderry. During that hour the hospital cases were taken into the ambulances, the remainder were put into buses, the two P.O.'s who had been drunk on the Sunday night and had been under open arrest ever since were taken away under escort from HMS *Hecla*, and our bread and meat that we had ordered by signal were embarked. I went ashore to give parting encouragement to the injured in the ambulance, and met an old school friend, David Colville, Lt, RNVR. That same evening we were back at our base in Londonderry.

To complete the story – we learned that the fire in *Comorin* was an accident. A broken oil pipe at the top of the boiler room had dripped hot oil on a Stoker and made him drop a torch which had ignited the oil on the deck. In a flash the boiler room was blazing and had to be closed down at once. This meant that there was no power on the fire main and therefore nothing with which to fight the fire.

Lincoln came into Londonderry two days later and told us of the sinking of *Comorin*. They had fired sixty-three rounds and two torpedoes at her. The torpedoes – not supposed to be fired at all being American pattern – had not hit. They did not reckon at first that their shells did much good. But as the fore hatch cover was near the water, owing to her pronounced list they pounded this and made a hole there large enough to flood the forehold. One of the motor boats floated off undamaged. Then the ship heeled over and finally the stern came right out of the water and she plunged down. She had sunk by about noon Monday 7 April. *Lincoln* had saved 121, *Glenartney* about 109. *Lincoln* had most of the army survivors, one of whom had been crushed between boat and destroyer. There was a story that an army officer after embarking in the *Comorin*'s boat had shouted up, 'Of course you'll be sending the boat back for our baggage!'

From *The Eye of the Wind*

32

A Battle – and a Collision

Now that SGB6 was ready for sea, there began a grim summer of fighting in the Channel, perhaps the most hazardous and certainly the most nerve-shredding part of my life. I now found myself often making mental reservations about the future – for the first time seriously allowing the possibility that I might not survive the war. I believed only thus far in life after death: that if I were killed some of my works – my own creations, pictures, books – might live on a few years after me, that the love of living people would do the same, and that my child and her descendants would move and talk and feel a little like me long after I was dead. But this feeling of continuation of my line was small compensation for the awful partings with my wife. By what right had I blighted her life with this perpetual anxiety? How could I justify myself in acquiring these new family responsibilities when I knew that I must go on fighting my war all out? It was the only way I knew how to do it; perhaps the only way I knew how to do anything.

On the evening of 9 April 1943 I found myself once more on the bridge of a Steam Gunboat, leading three small MTBs out of Newhaven harbour on an offensive sweep along the enemy coast. That night, I remember, there was exceptionally brilliant phosphorescence in the sea, so that the boats made beautiful luminous green bow waves as they sped across towards France. This would have seriously affected our chances of a surprise attack had enemy shipping put in an appearance, but they didn't. As we lay stopped and waiting we were apparently picked up by the German radar, which was very much less effective than our own. Nevertheless the shore batteries opened a fairly accurate fire on us, which made us move on. Later, at a range of at least seven and a half miles, they started up again and put a salvo of four shells very close to us indeed.

Four days later we were halfway across the Channel listening to the nine o'clock news from the BBC on our way to the German convoy lane when a shattering announcement came over the air. 'Hitch' had been killed in action off the Dutch coast. Robert Hichens, my old friend of fourteen-foot dinghy days, the Falmouth solicitor who had won more decorations than anyone else in Coastal Forces, had been struck and killed instantly by a stray shell right at the end of a successful battle. I remember that we stopped, as was our wont, to compare positions at the entrance to the swept channel through the minefield, and at the same time passed the news to other boats. There was a shocked incredulity in their tone as they answered. Surely there must be some mistake, they seemed to say. Others could be killed in action, but not Hitch.

It was another blank night with no sign of enemy shipping. I spent most of it thinking of Robert, and the day he had first appeared with the home-made fourteen-footer which we had all been snooty about until it came third in a twenty-knot breeze. I remembered how he had started just in front of us in the Prince of Wales's Cup at Falmouth when John and I had first tried out the trapeze and how in his surprise he sailed off the wind just long enough to give us our wind free and let us through. I remembered how he had taken me to sea on my first visit to HMS *Beehive*, the Coastal Forces base at Felixstowe a year before; how we had met only a week or two before in London to plan a new assault on the Admiralty in order to get our policies on heavier armament through; how I had heard that he would very shortly be asked to take on a training job at HMS *Bee* at Weymouth as a rest from operations, and how I had greatly doubted if he would accept it.

For me it was a cheerless, empty night.

Three nights after Robert was killed I took SGB6 and two D-class MGBs to patrol in the Baie de la Seine. One of these MGBs was commanded by John Hodder, in whose previous boat at Dover I had had my first introduction to the Coastal Forces. The other was commanded by Dickie Ball.

In bright moonlight we fought a lively battle against a defensive patrol of armed trawlers. The details of the story I sent in a letter to Jock Ritchie, the Captain of SGB4, soon afterwards.

> You were quite right about the brightness of the night. It was virtually a daylight action. Fortunately, the Hun shooting was very wild, partly because they thought we were a destroyer and were correspondingly scared, and partly because – well, after all, they were only old trawlers stooging along. We sighted them at three and a half miles. Unfortunately they turned 180° just then, otherwise we might have been able to sneak up from astern. I made a balls because Sandy [Sandy Bown, First Lieutenant and Gunnery Officer of SGB6] kept repeating (quite rightly) 'three-inch gun won't bear', so I turned to open 'A arcs' much too soon. The trawlers turned away and the result was 600 yards. I tried to get back to starboard but the steering had gone. The reason for this has not yet been traced as we can find no hits affecting it.
>
> The action lasted eight minutes – the last four of which only the two Ds were firing as I reckoned the range was too great. When we stopped firing (we had been using tracer owing to the brightness of the night) the enemy concentrated on the two Ds and left us alone. They had hit us a few times and killed the loading number of the three-inch gun. We hit them fairly hard – the second ship principally and also the third. Two or three times there

were sudden 'flare-ups' on the gun platform of the second ship, as of small cordite fires. Tracer was very blinding and so bright that the moonlight seemed to make no difference to it. The targets were almost directly up moon for all the first part of the action.

We tried steering by engines but the rudders were at Port 20 so it was hopeless. We could see the Germans about 2,000 yards away flashing to each other. I passed a signal to John Hodder and Dickie Ball 'Attack again' and they went off. Meanwhile, I managed to get the ship pointing well to starboard of the course and then rang down half ahead port and in this way progressed slowly in the right direction, stopping when the course did not help and turning to port with main engines, thus describing a series of loops.

Meanwhile, Ball in 615 saw me coming and thinking very reasonably that it would be nicer to attack with an SGB than without, he turned and closed me, leaving John proceeding at slow speed towards the enemy and already quite close (1,200 yards) but not being fired at. Well, I did not think much of him going in alone so I called him up and made 'Rejoin'. The signalling was very good all the time and it was clear that until my steering was right, there was nothing much to be done. Dickie Ball came alongside me, went astern and nearly rammed me stern first but just went ahead in time (two feet, but a miss is as good as a mile). He told me by hailer (which I was afraid the Germans would hear) that he only had four rounds of six-pounder left and a shell jammed in the barrel of his pom-pom. I reckoned he was a liability and told him to withdraw to our rendezvous five miles north-east of the action, wait there half an hour, and then return to harbour independently. This was undoubtedly an error. He still had a single Oerlikon and his point-fives and I reckon he would have been most useful if only to draw the enemy's fire. After all, many an MGB has had to fight with less than an Oerlikon and two twin half-inch machine guns.

We didn't get our steering right for half an hour, indeed it was quite a relief when we did. We set off, in hand steering, to get down moon again before attempting to intercept. The chap who came over from *Excellent* to investigate our gunnery shortcomings, put in his report that too much importance was attached to the moon aspect. That is actually a rather inept criticism, I think. My own view, and that of everyone who was there, is that had we approached from the up moon side, we should have been at a grave disadvantage. The trawlers could *see* us down moon, but they couldn't see us well enough to aim straight. I am convinced that they fired substantially more bullets at us than we did at them and the reason they were hit and we were not was very largely, if not entirely, because we kept them against the moon.

Well, we beetled off eastward at twenty knots with John Hodder astern and then after half an hour (rather less) we turned south, thinking we were ahead of them by then, and started to sweep back at slow speed. I had a fear, however, that they might have gone more to the southward, so we turned back to south-south-west and suddenly sighted them right ahead about three and a half miles with terrific bow waves, obviously legging it at speed.

I must own to a sinking feeling on sighting. The moon was so bloody bright. Whilst there was a doubt about finding them, the cowardly subconscious was saying hopefully, 'Perhaps you won't', although the conscious went on working out every possibility to make sure we did find them.

The leading trawler opened fire on us when we were turning towards and were still about 1,200–1,400 yards away. I do not think it possible to stop guns crews from ducking (at any rate, not until they are veterans, and this was their first action) if they have to be shot at without replying, so rather earlier than perhaps was advisable, I gave the order to open fire. The range was closed to about 600 yards and then it was quite obvious that we could not turn far enough to starboard to cross the leader's bow. I was not at all anxious at that stage to swop punches on parallel courses at close range as the enemy was firing four-inch, at least one if not two two-pounder, and at least four Oerlikons. So I turned away to port, and checked fire so as not to waste ammo. This, more by chance than design, worked a treat, as the Germans went on firing away with very little success as we turned to port and set off to the eastward to regain our lost bearing. (I reckon they were doing something like twelve knots.) We got a bit of a pasting as we got in line with 608 on the turn, as the overs (and everything was high) were hitting us.

But the point was that the Hun pans were good and empty when we were ready to turn hard-a-port and go straight for him with the *full advantage of the moon*. Up to this point I think (mainly by chance as I say) that we had adopted the best tactics for getting to close range under what were virtually daylight conditions.

We opened fire at about 500 yards at Red 30 degrees and by the time the leading trawler was abeam it had been silenced. The moment of realisation of this was extremely exciting. I turned to starboard to come in again, and 608 hauled a little out of line to starboard because she thought the trawler was trying to ram her as it was turning to port. I think this was a bit far-fetched as I am convinced the trawler was unable to steer by then and was rapidly losing way, with a loud noise of escaping steam. The result of 608 hauling out of line provided an interesting exercise in gunnery control. For a short while 608 fouled the range. Fire was checked easily for a few seconds and reopened again as soon as the target was clear of 608 – a matter of no more than ten seconds. This was possibly largely because of the brightness of the night and Sandy's very good handling of the guns.

Now I checked the fire and tried to get the three-inch gun to carry on firing, but communications to it had failed. After this it only fired two more rounds before it broke down for good. The semi-automatic gear was not properly disengaged and the gun ran out and would not come forward again, as it had bent the rod. We did one more run by which time practically all ready-use ammo was expended.

It was now getting very late and it seemed to me to be too late for a boarding operation, particularly as the trawler had manned a .303 machine gun and was evidently not quite ripe for boarding. She was spraying us with inaccurate bursts. I therefore decided to fire a torpedo and drew off to about 500 yards on the down moon side. The trawler had now swung back to starboard and I reckoned it had stopped. The first torpedo to report 'ready' was the starboard one and so that was the one I fired – at a range of about 400 yards or possibly a little less.

It is a remarkable feeling as soon as you have fired, isn't it? The awful irrevocability of it and the rather abandoned feeling – 'Well, for better or worse, that's done.'

The torpedo ran all right as far as I could see but the track passed about two or three yards astern of the target. Then I realised that I had been a complete ninny to assume that she was stopped, as being a very big trawler she would be bound to be carrying a little way still. Actually the torpedo may have passed under as the range was probably less than I thought.

Anyhow, I was so horrified by missing that I decided (quite incorrectly) not to fire the other one. Instead we ranged up alongside the trawler at about 100–150 yards and fired at her with all that would still fire – which turned out to be the starboard point-five and nothing else. Pom-pom supply had broken down owing to one yellow supply number; Oerlikon pans were empty and so were .303 pans, and the three-inch, which owing to the breakdown in communications we were expecting to open fire at any moment, was irreparably bent.

The point-five was using no tracer and was raking the target from end to end. You could hear the clatter of it against the trawler. But suddenly, the after Oerlikon on the target came to life and its first burst went into our bridge and into the caboose below. One incendiary shell came into the bridge and exploded knocking us all down (five of us) but only wounding Harris (the navigator) and him only slightly. The bridge was full of smoke and luminous incendiary composition and splinters of the binnacle and there was a goodly fire going in the caboose amongst the point-five ammunition boxes. I had been going slow both, but decided that I had better disengage until the fire had been dealt with so I drew off about 1,500 yards.

Meanwhile, John Hodder had silenced the German fire with a beautifully placed burst of Oerlikon at 200 yards range. He followed me and came up alongside, reporting that his pom-pom was out of action, he had used his entire outfit of six-pounder and that one of his twin Oerlikons was also out. This left him with one Oerlikon and his point-fives.

Our own position, when the fire had been put out, was that we had no ready-use ammo left for anything. Sandy said it would take about half an hour to be ready again. By this time it was pretty late and I didn't think we could afford to wait half an hour. Actually, of course, it should have been possible to be ready in less than that.

This is a part of the proceedings about which I am *not* very proud. We ought to have had another go. If there had been any guns to fire we would have gone in again at once, but with what I took to be an inevitable delay of half an hour, and with John Hodder's guns more or less finished, I decided to go home – without even trying a depth charge. Rather a poor show. We had a long way to go and as it was, we were pretty late.

During the last part of the action we were struggling on the bridge with coils of wire, as the aerial had been shot down. However we rigged a jury aerial and made a signal to Ball (615) to return to harbour, in case he should be hanging about waiting for us. The Germans switched on Cap d'Antifer light so as to give us a nice departure and we sped homeward. Our side were less thoughtful about navigation and gave us no RDF [radar] position. Nor did they give us any information about 615, although they must have heard that we got no answer to our call – and should have realised that we should be anxious. She must have been on the Portsmouth Plot. But then Duty Commanders never did have much imagination.

Since that party we had a rather bad do last week (night of 29/30 April). We went to cover a minesweeping operation which was cancelled owing to the darkness of the night. We went on and did a sweep between Fecamp and St Valery – four boats – SGB6, MGBs 608 (John Hodder), 614 (Peter Mason) and 615 (Dickie Ball).

It began to blow up so I decided to start back in good time but we had not gone far when I lost sight of 608 astern. It was shocking weather – wind force six and rain – so I switched on my overtaking light. John was so taken aback by the bright light that he stopped, thinking that we had either stopped too, or dropped a flare over the side. I went on, reducing to twelve knots for a mile or so and then, supposing them to have had a breakdown I turned back (a mistake), reduced to ten and flashed all round ahead. Presently, I picked up 608 on the starboard bow and signalled to her. She passed on opposite course about 150 yards away and 614 (Peter Mason) astern of her and out on her starboard quarter passed very close. 615 (Dickie Ball) passed closer still – fifteen to twenty feet. All very hair raising. While this was happening there was a splintering crash astern and 614 took 608 amidships on the starboard side. 608 had been turning to starboard to follow me. 614 had been on her quarter because of the AIX gear on her pom-pom which obscures the view ahead.

This was 0438 and the position was twelve miles from St Valery, i.e., in full view when dawn should break. 608 had a huge hole in her engine room and seemed to be sinking. 614 was hardly damaged. In the dark it was extremely hard to make contact with all the boats and it took ages to find 608, but eventually I got alongside and had her in tow by 0515 at seven knots but it was too ambitious in that swell, and after fifteen minutes it parted.

I was in a desperate quandary about W/T silence. A signal pin-pointed by the enemy at that time would be bound to stir up trouble. The weather was bad for aircraft – low cloud and rain. On the other hand, if I made nothing and the weather cleared it would take ages longer to get air cover. This was the sort of decision which might well have been made by the spin of a coin, but when seventy men's lives may depend on it, there is real agony of mind.

When we got 608 in tow again I finally decided on a signal and made it. Ten minutes later the tow parted again. By this time our towing pendant was in poor condition and I told Dickie Ball (615) to take over. It was now daylight and thank God the coast was hidden by rain. Dickie did a very good job. He got 608 in tow in a very short time and we had to accept a speed of four knots. However, the weather seemed so impossible for aircraft that we breathed more freely. Meanwhile, 608 got lower and lower in the water. 615 passed pumps and hose astern on heaving-lines and generally displayed considerable initiative.

We fell out one watch and I went down to the bunk in the charthouse for a few minutes but at 0730 shot on to the bridge when RDF reported aircraft bearing Red 120 and soon after I got up there a Ju 88 came out of the cloud and circled to the west of us where the horizon had suddenly cleared. The clouds were still low and the Hun turned up into them. He came out of the cloud heading straight for us and the three-inch was told to open fire. The Ju 88 crossed from port to starboard quarter and the three-inch gun was spoiled by the after depression rail. When they got back on to him he had turned steeply up towards the clouds and a moment later had disappeared. No shot was fired.

The next two hours with visibility ten miles and cloud base about 1,000 feet were fairly nerve-racking – but nothing came of it. By 0930 the rain had come down again, and we could breathe freely once more. At 1100 two MLs joined us and at 1200 we got a signal to say fighter escort could now be expected, but this was cancelled at 1230.

The sea got calmer in the afternoon and we got into thick fog off our own coast. They gave us a few plots and we found that our dead reckoning was only one and three-quarter miles wrong since 2300 the night before, which considering the towing and messing about we had done, was pretty good.

Three miles off Newhaven 615's tow parted. We had decided to let him finish the job, so he went alongside and took 608 in, berthing her, without assistance, at 1605. John Hodder's ship's company had been at the pumps continuously for eleven and a half hours. Really Ball's seamanship was excellent, but apart from that it was all rather a black show.

Principal causes of the collision were:

1. Darkness of night – very exceptionally dark.
2. 614 being on quarter instead of astern of 608, because of the special fitting on the gun which obscured the view dead ahead.
3. 614 altering to starboard, when she lost 608's stern light, instead of to port.
4. SGB6 switching on the overtaking light in the first place.
5. 608 misunderstanding it.
6. SGB6 going back to look for MGBs.

Board of Enquiry is on Friday!

From *The Eye of the Wind*

33

Dawn off Cherbourg

The refurbished *Grey Goose* was not finally ready for sea till mid-June 1943, and then we were sent off to HMS *Bee* at Weymouth to get 'worked up' once more into a fighting team. On the staff there now was my friend Jock Ritchie, who had had to leave SGB4 because of an old wound in his knee. His influence on the training programme was already apparent. A month later we were back at Newhaven to join up with four other Steam Gunboats. For the first time since I had taken over the flotilla a unit could be mustered for offensive operations which consisted exclusively of SGBs. The patrols began again two or three nights a week, but there were only minor adventures, until on the early morning of 27 July we were involved in a fairly desperate battle. At this time we were trying to co-ordinate the night bombing efforts of the old Albacore biplanes with our own attacks, and the shore-based radar on the cliff tops of the Isle of Wight were able, on nights when conditions were good, to give us some details of the movements of enemy shipping though they were not always accurate and often incomplete. The Cherbourg Peninsula was just within their range but the Seine Bay was outside it. When they detected shipping the details were passed to us by C.-in-C. Portsmouth and were called 'Enemy plots'.

On this particular night we had left Newhaven five strong, and in order to understand the story a brief *dramatis personae* is necessary. As we steamed south across the Channel, I was leading in *Grey Goose*. I used always to try to sleep on the way over, so that until the next alteration of course my First Lieutenant, Jock Henderson – an indomitable Scot – was, no doubt, keeping watch. Immediately astern was *Grey Shark* with Howard Bradford now back again in command. His First Lieutenant was Sandy Bown, who had served me so well during the brief weeks when I had been commanding her, and had done such an outstanding job controlling the guns in our Seine Bay battle. Astern of *Shark* came Grif in SGB8, now *Grey Wolf*, with an admirable First Lieutenant in Tim Langridge. Then came *Grey Owl* with Richard Hall, and finally young Jimmy Southcott, who had taken over SGB3 from George Pennell; she was now *Grey Seal* with Mike Barrett as First Lieutenant. This was the proud force that set out to intercept an important merchant ship which was expected to cross from Cherbourg to Le Havre during the night. It was soon to be reduced by one when *Grey Owl* developed a defective feed pump and I had to send her home. We drew blank in the Seine Bay, but went after a patrol – which turned out to be rather a strong one – just outside Cherbourg as dawn was breaking. Here is the letter I wrote to Jock Ritchie a fortnight afterwards:

My dear Jock,

I dare say you'd be interested to have the inside story of our do. I already seem to have written about fifty pages on the subject, but an action report – even with all its enclosures – is never more than about half the story. The only good thing about the affair was that we learnt an awful lot. If our experience can be used then we shall have gained something, I suppose.

To begin with, our patrol was designed to meet a merchant ship and not to concern itself with the R-boat patrol off Cherbourg. However, since our co-operating Albacores had swept the Baie de la Seine (we supposed) and found nothing, we decided that we should make all speed for the enemy plot off Barfleur.

Our Albacore boys had spent the whole of the afternoon before with us – and that was their second visit to Newhaven. We had made what we thought were fairly watertight plans for identification; furthermore if they met the enemy they were to illuminate after bombing. Somewhere about 0200 they did bomb the R-boats off Cherbourg and illuminated them, but they were round the corner from us and still about thirty miles off so that we didn't see the flares. If we had it would have saved the twenty minutes delay in getting the first enemy report out to us.

Portsmouth gave us two enemy plots which were quite good but the third and last one was phoney I think. Anyway it looked like a right-angle alteration of enemy course from west to north. We didn't get to the position until nearly an hour later but received no more plots although I believe the enemy were being plotted all the time. This resulted in our slowing down much earlier than necessary and finally, after passing through the positions, stopping to discuss the situation. Grif came up to starboard. We were north of Cap Levi and if we were to go back the way we had been told to go (010° from Barfleur) we could not afford to go further west. On the other hand, the enemy was clearly still at sea, and probably heading for home. If we went westward after him, we could always go home to the north-west if we missed him, and we could go on hunting for another fifteen minutes. Grif said, 'I think it's hopeless now without another plot. I think we ought to call it a day.' I said, 'We'll go on towards Cherbourg for ten minutes anyway before we give up.' It was one hour and twenty minutes since the last plot had been received.

By this time the moon – which was after the last quarter, not much more than a crescent – was well up and giving a little dim light, but not much. It was almost directly behind us. We cracked on again at twenty knots, course about 240° and a few minutes later I suddenly heard four explosions astern. I had seen no flashes but nevertheless assumed it must be shore batteries as we were about four miles off Cap Levi where we knew there was a radar battery. So I immediately altered course to about 330° to open the range. Actually it was an Albacore which had straddled Jimmy, who was tail-arse Charlie, but I did not discover this until we got home. The Albacore was from Exeter, and not one of our friends from Tangmere. It had been allowed by its controller to come as far east as Cherbourg and thought it was off Alderney. Actually, it was due north of Cap Levi, four and a half miles. Two bombs fell neatly to port and two to starboard of poor Jimmy and a burst of tracer passed across his bridge.

As I steadied on the new course I got a radar contact at Red 80 [this means 80° on the port bow – that is to say, almost on the beam] 3,500 yards, two or three small echoes. So we immediately came back to port to bring them ahead. I asked Ian what the time was and he told me 0355 to which I am alleged to have replied 'We've just got time for a quarter of an hour's fun and games.' It was within five minutes of the last possible time I reckoned we could stay with safety from air attack on the way home.

Note: the *correct* thing to have done from the point of view of the war effort and material would have been to have turned north at once and returned to Newhaven. From a morale point of view it would no doubt have done harm and I expect I should have been court-martialled, but I am sure it would have been a most sound and sensible thing to do – as we knew we had only fifteen minutes and we knew that a good attack takes an hour at least. We also knew that the patrol was a strong one and that it would be there another night. So it would have been right to wait until conditions suited *us*. However we didn't.

We didn't know exactly how close to the shore we were. I thought the shore batteries were already firing and I did not fancy being hemmed in by a strong force and so when I saw the Huns ahead I turned slightly to starboard so as to engage them on the port side firing to southward. A turn to port to go inside them would almost certainly have been better as it turned out.

There seemed to be no end to the Germans. They looked like a flock of wigeon with two or three geese amongst them. The sea was so glassy that you could see the reflections of stars. The moon was nearly behind us – slightly on the port quarter. Also behind us was a faint north-east glow of the dawn. The geese were on the left – two or three trawlers – and the wigeon were sitting still apparently, and all clustered together. Being a punt gunner I aimed for the thickest part of them – not, of course, a very good idea unless I was going to use my 'punt gun' (torpedoes) which I wasn't. Actually, of course, I should have, as it was a good chance of a flock shot – if they had been set to three feet. About three or four of them in the middle were quite solid. In fact at first I thought there was a longer target there – but there wasn't. There were far too many Huns to count – at least eleven.

At about 500 yards the radar reported: 'Echoes lost in the ground wave, sir.' One of the R-boats challenged with quite a dim blue light – I'm afraid I couldn't read the letter – ?R, ?F. We came to the trawlers first, and I let them go past so that the other boats would have something to shoot at when we opened up. I think the line was a bit longer than it should have been just then. (The order was Howard, Grif, Jimmy.) *Closer station keeping.*

When it is very calm it is hard to judge range but we were certainly not more than 300 yards and possibly a

> good bit less when I said 'Open fire'. Then a tragedy took place. The port point-five gunner lost his head and fired his guns without aiming, about twenty degrees into the air. This was blinding to the other guns and from that moment all hell was let loose.

Here I must break into the narrative to give some idea of what the scene looked like. I hardly needed to describe it to Jock, for he had seen this kind of thing more often than I had. But assuredly of all the actions I fought in the Coastal Forces, this one produced the greatest volume of tracer shells and bullets. At least fifteen ships had opened fire simultaneously and the air was thick with the red and green and white streaks. They ripped away from our guns towards the enemy line, they fanned out of the enemy ships and came lobbing towards us almost as though you could reach out and catch them, until they whipped past just over our heads singing as they went; they criss-crossed ahead and astern, they ricocheted off the water and popped as they exploded in the air, and they thumped into us from time to time with a shower of sparks. It would all have been very beautiful as a spectacle if it had not been so dreadfully frightening. But for me, thank goodness, there was no time at this stage to be frightened. The letter goes on:

> I'm afraid the Albacores' bombs – only 4,000 yards away – had fully aroused the Germans and every gun must have been trained on us. I suppose they were blinded by tracer too, because considering how close we all were to each other it is odd that more damage was not done on both sides.
>
> The Germans must have got under way fairly quickly because by the light of sudden flashes which I took to be hits – but which may have been gun flashes (I think they were hits though) – I could see the R-boats creaming along on a parallel course about 300 yards away. At this stage it would have been nice to have had some starshell.

Our three-inch had been doing very well and I saw three hits with it, which caused big showers of sparks. It would have been quite impossible to convey any speech to the guns – the only way would have been to check fire altogether and I was not keen on doing that. So we had no starshell. As a matter of fact, I don't think I thought of it myself till afterwards. In any case far too many R-boats seemed to be shooting at us and my gunners were not firing at more than two of them at most. I looked back and could not see any SGBs following. In truth I could see very little at all as the tracer was completely blinding.

I remember seeing the leading R-boat on our beam and shooting at us and thinking, 'Thank God there aren't any more to pass.' We had passed at least six overtaking at about five knots, and they had done us a bit of no good. We were burning a bit aft – with a lot of smoke; and a scare report came to the bridge. 'All the three-inch gun's crew killed, and we are heavily on fire.' We kept on a steady course still because three guns were still firing quite well, it seemed. The first aft died down, but then suddenly flared up very bright again and all the Huns started shooting at us so I decided disengage. A long way astern and on the starboard quarter I could see two SGBs flashing on their recognition signals (two green). As I turned away more and more ships seemed to be shooting at us, some from the starboard quarter and fancying that this must be the SGBs I switched on my display, too. It lasted about ten seconds and then – pink – and out it went, shot out by Grif or Jimmy or both. However, they only got two pom-pom hits on us. We found the fuses afterwards and neither of those did any damage to speak of – one went into the mess deck and one through a splinter mat and into X gun ready-use locker where it spoilt some pom-pom ammo but didn't explode it. I found afterwards that my starboard sided guns had fired about 200 rounds back. Shame on us!

Jock Henderson who was wounded and Able Seaman Wendon and Stoker Clelland put the fire out very quickly with a hail of bullets whistling round their ears. It was a three-inch ready-use locker full of starshell which had gone up.

Also an entire ready-use locker full of Z guns and pom-pom ammo had vanished in one 'pooff' and left the locker looking like a tulip.

I reduced to ten knots as soon as the worst of it was over in order to let the others catch up. From time to time some hopeful German would direct a burst at us and they would come lobbing towards us. I planned to go to the standard rendezvous two miles north of the action and see if we could go in again.

As I looked back I saw a line of all three SGBs in the light of some starshell which the Hun had just put up.

I watched them go in to attack and suddenly saw one ship get the most terrible pounding. I hoped it was one of them but I feared it was one of the Steamers because it was very near where I had seen them in the starshell. For about twenty seconds shells were bursting and sparkling all over it. The battle roared on very fiercely and then rather suddenly ceased. There was a pause of a couple of minutes and then the distress signal – a short burst of tracer going vertically upwards directly astern of us. I said to the coxswain, 'Starboard thirty – steady on south', and my heart sank into my boots!

Now to go back to the others . . . The action had begun at 0357 and early on *Grey Shark*'s steering had gone. If you remember it went in a previous action in a rather unexplained manner as the pipes were found to be slightly dished by a shell which was believed to have hit in the second phase of the action, *after* the steering failure. Again, although the pipes were well and truly shot away this time, there is some reason to believe that they were not shot away until after the failure. The mystery is still unsolved, but still being investigated. Anyway the effect was that, before it was definitely established that the steering had gone and the wheel could be changed from power to hand steering, the ship had turned nearly ninety degrees to starboard and contact with the leader had been lost. Grif did not know whether to follow Howard or not, as he couldn't see me – indeed he couldn't see anything much and nor could anyone else, so completely blinding was the tracer. So the line was broken and for a bit Grif and Jimmy took on principally the trawlers whilst I was taking on the best part of the line of R-boats up front.

But as soon as Howard's steering was fixed up he saw me burning brightly and decided that I must be in a very poor way and must be protected from the rough Germans. So without a thought he plunged into the mêlée between me and the R-boats announcing to Sandy his intention to ram. He missed, unfortunately (?), and passed about 25 yards from four of them, getting a horrible pasting as I had seen. He managed, however, to give pretty fair return, I gather and saw a lot of hits on the enemy. They also (Sandy and some of the guns' crews) saw two men in the water clinging to some wreckage. They wondered if they could be chaps blown overboard or washed off us (as they quite thought that we had sunk by now – and so did the Germans according to their communiqués). They also say they saw a ship very low in the water which looked as though it was sinking.

At that very close range some of the enemy's armour-piercing shells were bound to go through; in fact seven entered the boiler and with a hiss *Grey Shark* came to a standstill. The boiler room crew escaped unhurt either by the shells or the escaping steam, which is quite remarkable.

Grif was too blinded by the tracer and starshell even to observe that Howard had been stopped. All he knew was that he couldn't see the next ahead and that he was being shot at by ships at Red 30 and Red 130 and at every ten degrees in between. He could still see me burning on his starboard side and decided to follow round to starboard. His bridge had been hit, his First Lieutenant and navigator and coxswain all slightly wounded and he was at the wheel and coming round to north. The coxswain had been cracked on the head by a large piece of Perspex windscreen dislodged by a two-pounder shell on the bridge armour. He picked himself up from the deck and took the wheel again and went a little past the course, altering back to port.

Meanwhile Jimmy had seen Howard buy it. He stood on for a few seconds and then said to himself, 'Oi, you can't leave old Howard there,' and turned back hard-a-starboard.

Halfway round his turn he suddenly saw Grif on his starboard bow and as Grif altered back to port he suddenly saw Jimmy. There was nothing to be done. They both went full astern but they hit pretty hard. Grif's forward mess deck was holed very badly – a hole about fifteen feet wide at the upper deck going down to a point about three feet below the waterline. Fortunately the bulkhead between that and the old officers' space was undamaged, and so although he thought at first he was going to sink, he was in fact quite well able to keep the damage under control.

All those things had happened pretty quickly while we had been putting out our four fires (two forward and two aft) – and before Sandy had pointed his bridge Vickers guns into the air and fired a burst to summon assistance. We turned at once and steered south towards the distress signal and then in the binoculars I saw *four* ships lying more or less stopped, and a dozen possibilities went through my mind. Were they all Huns? Were three of them Huns mustered round the damaged SGB? The fire parties went back to their guns – but I must confess I wondered how many of them would still fire all right. Only two of the three-inch guns' crews had been laid out. I was a bit worried too because our recognition display had been shot away (as well as our W/T and radar aerials). Identity was the key to this situation and the best and quickest method was not available to us.

Certainly one of the four was an R-boat – he came more or less straight towards us. The others were directly behind him – and which were they? The next one looked all wrong. Surely another German. Actually it was Grif down by the head after the collision, about which of course I knew nothing.

Now came a regrettable moment. The first boat had been identified as an R-boat. It looked exactly like the photograph we have in our intelligence room. It altered course to starboard and passed close down our port side. It should have been blown out of the water. But it wasn't. Ever since I have been wondering just *why* it wasn't. The answer is a combination of reasons I suppose. I was still not sure that two more R-boats were not trying to board the damaged SGB; I did not know exactly where the remainder of the enemy were, but I did know from the distress signal that one SGB was fairly certainly going to have to be towed. It was getting light pretty fast and any interruption of the towing would probably spell disaster. A fresh battle would produce shore starshell and the towing operation would be detected. If the enemy knew what was afoot we should never be allowed to pull away our lame duckling. The other SGBs (with commendable loyalty!) stoutly maintain that the issue was a straight swop and that my decision was right. No doubt the R-boat in question is also quite happy about my decision. I am pretty well satisfied it was wrong, but I didn't have awfully long to make it in, and you know I'm mostly pretty slow about things.

By the time I had identified the other three boats as SGBs the R-boat was too far past us to make it worth starting anything. One of the SGBs got under way. I made 'Follow' to him by light, but although he gave an R for it, Grif didn't receive the signal and in any case he was best out of it. I, of course, did not know how badly damaged he was, until a little later when he made a W/T signal saying he was disengaging.

It appears that the R-boat had closed Jimmy whilst he was trying to get Howard in tow and started flashing, and Jimmy replied with E.B. ('Wait' in German signal procedure) and the R-boat had waited. Actually of course the R-boat must have had a nasty shock at finding himself so close to the enemy and must have been very glad of the opportunity to creep away. During this time *Shark* saw and heard a German shouting in the water about fifty yards away.

As soon as Grif had gone off I circled round the other two. I watched the R-boat circling to the north of us, and following round to the west with binoculars came suddenly upon a whole bunch of them, seven at least. The wretched things were clustered in the darkest part of the horizon whilst we were completely silhouetted against quite a bright dawn, and what was more, it rapidly became apparent that they were closing in at slow speed. And still the two SGBs lay stopped. I could distinguish, now, which was 'tower' and which was 'towee' and suddenly I saw the 'tower' start to pivot round: the strain was on the tow, they were under way.

By this time the enemy were less than 1,000 yards away and the obvious course was to lay smoke to cover the tow. I increased speed and started to make smoke and immediately the Hun saw what was happening and opened fire. However, I had already passed between them and the tow and the smoke had hidden it perfectly. The R-boats' fire was very poor and our B gun, which had had a stoppage in the first action was in particularly good form. This was Hitch's pom-pom gunner and he was firing in short but admirable bursts. We saw quite a number of hits on the second R-boat. *Seal* and *Shark* saw the glow of it through the smoke. The first R-boat turned towards and I expected him to come through the smoke after us but the second one turned away and the others seemed to follow. Some starshells were fired and several big splashes appeared about fifty yards away, either from the trawlers or from the shore batteries, probably from the trawlers.

Meanwhile the tow had whacked it up to about ten knots and Jimmy, with most excellent judgement, refused to attempt anything faster, realising that if the tow parted all would be up. We dropped astern of them again so as to be able to repeat the smoke run, and we did in fact make another puff of it for good measure. A few minutes later we ran into what we took to be a fog. Actually it was some smoke which Grif had made ten minutes earlier with the object of providing cover for us. 'Do you think it might come in useful to them?' he had said to the wounded Tim, who had just completed a big shoring-up job. And so the smoke had been made. I had been looking astern for signs of a pursuing enemy and when I looked round both *Seal* and *Shark* had disappeared. The sea was so calm that the smoke was completely invisible, and as we were still in the clear I could not for a moment imagine what had happened to the two ships which had been about half a cable away only a few seconds before. I asked the starboard gunners if they had sunk, but they seemed to think they hadn't and suggested fog, and then we ran into it too. In the middle of it we passed only five yards under *Grey Shark*'s stern. Very hair raising. After that we ran through several patches of real mist – so much so that I began to feel more sanguine about the risk of air attack.

By now it was broad daylight – I did not think the enemy ships would dare to follow long because of our air support which they would expect. We had a message by light from Sandy saying he was the only unwounded officer in *Shark*. Portsmouth made us signals about some destroyers which were coming to meet us and eventually we met the *Stevenstone* and soon after the *Bleasdale* with Grif in company.

We all stopped and, while *Seal* remained attached to *Shark*, I went alongside *Shark* and shunted her backwards alongside *Stevenstone* so that their casualties could be transferred – and ours also. This took about half an hour or more, in a position about halfway between St Catherine's Head and Barfleur.

Howard, though quite seriously wounded in the leg, was still hopping about on his bridge, Johnny Harris, his navigator, was pretty bad, with a compound fracture of the thigh, and is only just off the danger list now. His Midshipman (Tomkinson) died on the way back in *Stevenstone* and so did another of his chaps. Our Bill Williams-Ellis was badly wounded in both legs, but he's getting on all right now.

Howard's coxswain – himself badly wounded, amputated another man's leg (with a razor blade) in an attempt to save his life, but it was no good. The coxswain went on working till they got back when he collapsed.

Grif had already transferred his wounded to *Bleasdale*. One – the captain of the three-inch gun – had had his leg severed by a shell bursting on it. Supporting himself on the gun he had taken over breech worker and gone on firing the gun for six minutes until the action finished when he collapsed and died soon after – three others of Grif's crew were killed.

While *Shark*'s wounded were being transferred we went off about half a mile and tested our guns – all firing at once. This was a very good idea as we felt ready to tackle any F.-W. 190s that might turn up and the noise put heart into everyone.

There were seven killed and over thirty wounded and many of those thirty kept going, some of them for nine hours or so – till their ships were back in harbour – Howard did – and Tim Langridge – and my leading seaman and Howard's coxswain. Jock Henderson with a bullet in the back of his neck and blood pouring all over him, dashed about the place unceasingly. Once on the way home I looked up and saw him at the top of the mast in a bosun's chair trying to repair the radar aerial! At last his enthusiasm came to grief when we were detached to return at twenty-five knots and he set out to rig a bottom line. The chain carried away and wrapped itself round the screw and, presto! we were a docking job.

The whole do lasted only thirty-six minutes, but I reckon quite a variety of things took place in that time. There were a lot of opportunities for exceptional resource and initiative and courage and there were a lot of chaps who were not slow in taking them.

Jimmy's towing was the star turn – and Howard's headlong plunge.

The results? A lot learnt – some good experience, a few Germans killed, a few holes in some R-boats – and four SGBs out of action for periods varying between seven days and six weeks, with seven of their gallant men gone for good.

It appears that two R-boats had to go into dry dock and two more were not used after they had returned under their own power to Ouistreham. There is no indication about the trawlers. So far as is known nothing was sunk. One interesting thing is that at about 0436 the R-boats went back to Cherbourg, but just before entering one apparently turned back to the scene of the battle – possibly to look for the men seen by *Shark* in the water.

And the merchant ship we had been after moved from Cherbourg to Havre about five nights later!

The Germans say that two 'MTBs' were sunk. They say there were five of us and that one was set on fire and was seen to sink. Later during what they call a second patrol only three British boats were seen indicating that another had sunk. Very little return fire was encountered and they were able to close in without much danger to themselves. 'Piles of dead and wounded were observed on the decks of the British boats.'

Well, there's the tale. Another battle spoilt by bull-in-a-china-shop methods, but they were rather forced upon us by the time of day.

I would very much like to hear any comments you may think of. We have worked out a long list of lessons learnt. Some are fairly controversial, but we put them in so as to make people think. You will be getting a copy at *Bee* in due course, I expect.

Love to Phyllis and Jean.

Yours ever,
Peter

After that action I had two special worries. The first, which I could never quite escape in any battles where there were casualties, was the degree of my own responsibility for the deaths and injuries of my chaps. Perhaps if I had been more skilful, if I had turned to port instead of starboard at the beginning and had thereby got the advantage of the breaking dawn, perhaps if I had thought of using torpedoes at the outset and approached more slowly, perhaps if I had *recognised* that this was a sort of ambush we had fallen into and needed special cunning to turn to our advantage – so many perhapses. It was easy enough to talk about not making omelettes without breaking eggs, but what sort of an omelette had we made anyway?

And then, at the other end of the scale, was that wretched R-boat that got away. Down the years I have lain awake at nights and thought about it. Was it really just plain cowardice that prevented me from saying those two words 'Open fire'? Like a film editor I find myself winding back the film and starting it again at that moment when the R-boat detached itself from the other three. Clearly it had not seen us at all. It was heading almost directly towards us, so as to pass down our port side at no more than fifty yards range. All guns were given the bearing to train round on to. If they had all been on the top line and ready to fire, I should think it must all have been over in ten to fifteen seconds with the R-boat a shambles in front of us and no one left alive on the upper deck. But *were* all the guns on the top line? How many would have fired when I gave the order? Perhaps less than half of them. And then what about the three ships ahead, at least one of which looked like another R-boat? Perhaps two of them were at this moment trying to board and capture the disabled SGB. These were my friends who had made the distress signal and who counted on me to come to their rescue. There were still long seconds before the R-boat would be past, still long seconds in which to make my decision – my decision to say just two words. I was looking most of the time through binoculars at the dark shapes ahead, trying to identify them, and glancing occasionally at the R-boat. I was waiting for a report from the three-inch gun that it was 'on target' and ready to fire – a report which never came. Of course at that range neither ship could easily miss and certainly the R-boat's guns were trained on us. He may not at first have identified us, but as we drew level he must have known for certain that we were his enemy. Yet he did not open fire. Obviously he would not, for on the face of it we were the more powerful ship – larger and with heavier armament. If we didn't start anything, certainly he would not. But of course if we did, so would he.

No doubt we should get a pasting if as I feared we could not muster the fire-power to smother him in the first few seconds. Try as I will I cannot remember consciously thinking of this argument at all, but how large did it loom in my subconscious and influence the other thoughts that raced through my mind? Now the R-boat was exactly abeam passing on an opposite course, the ideal moment to open fire – one tremendous shattering burst from all our guns and the job would be done. But what about my friends in distress? I put up the glasses once more to look at them; two of the ships were certainly SGBs, even the third, odd though she looked, must be an SGB too. Now I must open fire of course. I turned back to the R-boat, which was on the quarter now – the best opportunity was past. A pity to start anything now that the best chance had gone. Better get on with the rescue of the damaged SGB which must be got in tow unmolested.

Paragraph 10 of the C.-in-C. Portsmouth's covering letter says:

> The Senior Officer, when he turned back to the distress signals of *Grey Shark*, was faced with a difficult decision when he found himself passing close to an enemy vessel. He could have done no wrong in engaging, and such would have traditionally been the correct course of action, but he was summoned by a consort in distress, and this might have made the difference between her loss or safety. As the battle eventuated, had *Grey Goose* been delayed, diverted or damaged by this potential encounter, she would not have been able to screen the damaged vessels at the critical moment.
>
> 11. The taking in tow of *Grey Shark* by *Grey Seal* in four minutes under fire shows a high degree of seamanship and training.
> 12. The action of the Senior Officer in laying a smoke screen between the concentrating enemy and the tow, and

> drawing the fire on himself, was a well judged and gallant action which met with the success that it deserved.

Soon after this battle I was awarded a bar to my DSC. But I still lie awake at night sometimes and wonder.

From *The Eye of the Wind*

34

The Story of SGB9

The Steam Gunboats – SGBs – had been designed as an answer to the E boat (the small German diesel-engined torpedo boats which attacked our Coastal Convoys in the Channel and North Sea); but unlike the Motor Gunboats and Motor Torpedo Boats which were made of wood, the SGBs were made of steel – steel in very thin sheets, because they had to be light in order to be fast; and again, unlike the MGBs and MTBs, they were powered by steam instead of petrol engines, which was thought to be more reliable and less inflammable in battle. Another difference was that they were quite a bit bigger. The small MGBs and MTBs were 70 feet long and some were even 110 feet long, but the SGBs were 150 feet long.

By the time that a flotilla of seven SGBs had been built, the job they were needed for had changed and so it was decided to put torpedo tubes onto them, and a lot more guns and use them for attack rather than defence.

I first saw SGB9 on a wintry day in March 1942. She was lying in the river Medina at Cowes alongside the shipbuilding yard. She was just a hull – no upper works, no guns. From astern she looked like a tin box, and a rusty and dirty one at that. It needed the eye of faith to see her as a finished, efficient, fighting ship, but that's what she had become by the late summer of that year, 1942. She was the last of the flotilla to be built and she was a pretty ship with a raked stern line, giving her a very pointed bow, a small squat funnel and even in those early days enough guns to look impressive. She was painted pale grey and pale green and white in a scheme of camouflage which I had invented myself, and she looked very smart and very slinky. She was fast too.

I remember hot summer days when we did our speed trials up and down the Solent along the measured mile in Stokes Bay – creaming along with a wide, flat wake behind us. She went 36 knots on those trials and we were very pleased with her, and then we took her off to Weymouth for what's called 'working-up' – a period of concentrated training – exercises, drills, lectures, and more exercises. We were impatient about this not because we didn't need the training, but because we had an idea that there was something in the wind – an important operation of some kind, and we were desperately afraid we were going to miss it. But we didn't – in fact we were ready with time to spare and there was a week's leave for the ship's company.

At that time our ship's company consisted of three officers and thirty-five ratings, and we all lived on board – in contrast to the smaller MTBs and MGBs whose crews lived ashore in the Coastal Forces Base.

On the evening of 18 August 1942 we slipped from our berth in Portsmouth Dockyard and followed a group of Infantry Assault Ships down the harbour to Spithead. We were off on SGB9's first operation. Our new guns were going to be fired for the first time in anger. We were part of the Dieppe Raid.

Dieppe was a sort of diversionary raid to stretch German resources and ease pressure on the Russian Front. Its strategy was to hold down large defensive forces by the threat of more raids of its kind, any one of which might turn into an invasion. It aimed to draw a lot of enemy aircraft into the sky so that the RAF could shoot them down, and it aimed to learn something about the German methods of defending their so-called Atlantic Wall.

In the event, it did all these things, but the cost was high, especially among the Canadian forces who were landed, and most people regarded it as a failure. Personally I think they were wrong and most historians of the war now realise, I believe, what it achieved, but that's all another story. We weren't concerned with the strategy, only with our particular jobs of escorting and protecting the bigger ships and covering the landing of No. 4 Commando with our guns.

It's twenty-five years since that night – and I find myself remembering it in a series of pictures – rather disjointed pictures. Maybe that's because I'm a painter. But it's harder to paint them in words.

The first picture comes early in the night. We were steaming south across the Channel at 19 knots. It was quite calm – no wind, no sea, bright starlight – and we kept our station on the beam of one of the troop-carrying Assault Ships. I remember that I had a canvas chair brought up onto the bridge and sat in it trying to get some sleep which I thought it might be handy to have before the action began. But you don't get much sleep at times like that. I lay back and gazed up at the stars which moved gently round the foremast rigging above me, and wondered how our little ship would acquit herself, and whether our training had been good enough – and whether we had a return ticket or not. But it was not much use to speculate on that; it was best to assume that whatever happened one's own ship would be all right.

My next picture comes at a quarter to four in the morning. We were steaming in slowly, escorting the eight little landing craft which had been lowered by our Assault Ship. Ahead we could see the winking light of the Pointe d'Ailly Lighthouse, and on the southerly breeze we could smell the hay fields of Normandy. Suddenly a starshell went up away to the east, and a fierce battle of tracer bullets flared up and then subsided. Part of the left flank of the Assault Force had run into an enemy patrol of Armed Trawlers, but still it seemed that the main force was not detected.

At the first light of dawn – 4.50 – was the time for the touch-down. Six minutes before that time our aircraft began their attack and white tinselly tracer went up from the guns round the lighthouse. Still the ships were undiscovered. About a minute before the touch-down a single white fireball lobbed up into the sky. Then the pillbox at the eastern end of our beach opened fire at low angle, and the fire was returned by one of the landing craft. The Germans behind the beach fired a six-star green firework – no doubt an invasion signal – and the party was on.

At a range of 1,000 yards we opened up with our 3-inch gun at the pillbox – and SGB9 had fired her first shot at the enemy. After about the sixth round the pillbox stopped firing. Later it started up again and we gave it a three-minute bombardment, after which it remained silent; we hoped we had helped the Commando boys to get safely ashore. We had to draw offshore a bit when a heavy gun started lobbing shells at us. And then when it was broad daylight we were joined by SGB8, and we set off on a sweep to the westward to give warning of the approach of some German destroyers which were known to be in Le Havre (but which in fact never turned up). We steamed slowly along the enemy coast as bold as brass. At this moment in the bright morning sunshine we were lulled into a most curious and entirely false sense of security.

There, four miles away on our port beam shone the cliffs and the bright green fields and woods. St Valery-en-Caux nestled in its hollow, with a haze of chimney smoke above it. Behind us – far behind us now – the battle raged round Dieppe, and the Spitfires circled in hordes to protect the shipping. Somehow we forgot that unprotected shipping outside the fighter umbrella would be exactly what the Germans were looking for. We didn't think of the watchers on the cliffs feverishly telephoning for the Luftwaffe to come at once and bomb the two unprotected ships to the westward.

I remember saying to my First Lieutenant, 'You'd better send someone to get some breakfast for the ship's company, No. 1.'

When the first enemy aircraft appeared (they were two little Focke-Wulf fighter-bombers) we suddenly realised our predicament. They went past without attacking, but ten minutes later two more came weaving towards us, and worked their way up into the sun. Almost at once we could see that they had designs on us, and as the first one turned his nose down I could see the bomb hanging underneath. 'Hard a-starboard' – and I rang up the revs to 28 knots. Off came the bomb just as our guns opened up. It fell in our wake close astern. Meanwhile we saw the second Focke-Wulf coming in on the port bow. As he steadied up towards us in a shallow dive I saw splashes in the water short of us from his machine guns, and then our own guns answered. I saw his bomb enter the water 20 yards short on our port beam. There was a pause and then a heavy shock and a huge waterspout – but the ship was still afloat and still steaming. I remember thinking it must have been a very small bomb not to have damaged us more. Then I looked at the Focke-Wulf. A trail of wispy black smoke was coming out of it, and it began losing height. When it was nearly down to the water it picked up again and began to climb. I stopped watching it and became concerned with the fact that the ship would not steer, and that the alarm bells were ringing continuously – in fact, that we were pretty badly shaken. But those who did watch the damaged aeroplane saw it falter again and crash head on into the bottom of the cliffs.

We had to stop for repairs, and SGB8 took us in tow, but our engine room staff, working in desperate heat because the fans had failed, soon got her going and we could steam back from our sweep at 30 knots, back to the protection of the Spitfire umbrella.

Throughout that day we helped to maintain the smoke screen which protected the shipping. We shot at a great many enemy aircraft, and on two occasions we plunged through the smoke to bombard the shore, in answer to radio signals asking for support. For these bombardments we were only about 600 yards from the shore, which in broad daylight *was* a bit hair-raising, but we were lucky and we weren't hit much.

During the day we picked up a good many parachute survivors, mostly from enemy aircraft but some from our own – one was a Norwegian and another an American – and we also picked up some of the crew of the Hunt Class destroyer *Berkeley*, the only destroyer to be sunk during the operation (she was hit by a bomb, and finally had to be sunk with torpedoes).

Then when the withdrawal began in the early afternoon, we found ourselves with various rather unpleasant jobs to do, each of which took us closer and closer to the shore; and now that all the smoke screens had dispersed we felt horribly naked. A near-miss bomb had blown five of a gun's crew off the destroyer *Calpe* and she asked us to go back and look for them. We found and collected all five of them, and were just setting off when we sighted a landing craft coming away from the shore. There was nothing for it but to go in and take it in tow. For some incredible reason we were allowed to do this without a shot being fired at us, although we were less than a mile from the beach.

I have one last picture of the Dieppe Raid. As we came away keeping company at 12 knots with a damaged Motor Launch – we were the last ships away by 5 miles – a Dornier bomber thought we were money for jam. He came from right astern and dived down as we increased speed. All our guns opened fire and I watched the bombs come out. At once I saw that they were travelling in the same direction at about the same forward speed, perhaps a little more. They would fall just ahead, so above the uproar of the guns I yelled to the Coxswain, 'Full astern both.' He didn't hear me the first time, but when he did the result was most striking. The ship pulled up dead in her tracks, and we were pitched forward on the bridge. The bombs went on to fall about 60 yards ahead of us. Meanwhile the guns had been doing well. The Dornier was hit by a burst from the 3-inch gun under its starboard engine, which caught fire and a thin stream of smoke came from it as the aircraft plunged almost vertically downward. There was great excitement on the bridge; 'We've got him, we've got him,' and everyone danced with delight. But when he was a couple of hundred feet up he flattened out and, still burning, disappeared into the haze over Dieppe.

And so we joined the convoy and returned safely to Portsmouth just after midnight.

That was our Steam Gunboat's baptism of fire – her first great day of action. But, of course, it wasn't the normal pattern. She was in a great many battles thereafter, but nearly all of them at night, fighting enemy surface ships – sometimes E boats, but more often the heavily armed escorts of German convoys creeping along the French coast under cover of darkness.

These were the targets for her torpedoes. For these purposes we found she needed some alterations. First of all she needed more guns – lots more. So we persuaded the Admiralty to plaster them all over her so that she fairly bristled. Then we found that her machinery was very vulnerable. Whereas the Motor Boats had three engines, some of them four, we only had one boiler and if it got hit we were done for and had to be towed away by our chums. So we put on a belt of heavy 1½-inch armour plate all round the machinery space.

Originally our displacement had been about 160 tons, but now it had gone suddenly up to 240 tons, which is quite an increase. So, of course, we couldn't go so fast. Our speed came down from 36 knots to about 28, but it didn't seem to matter so very much because all our attacks were creepy crawly – sliding in towards the enemy at 8 or 10 knots, so as not to be seen. If you go fast at night your bow wave shows up, and if you're seen the target can alter course and avoid your torpedoes, whereas if you're 'unobserved' (that was the official term) you have a much better chance of success. All the loss of speed really meant was that we couldn't get away quite so quickly after the torpedoes had been fired.

While these alterations were being made I became Senior Officer of the Flotilla of six boats (the seventh one had been sunk some months before) and that meant that our SGB9 became the Flotilla Leader, and had to carry a rather larger ship's company, and, therefore, became much more crowded and much less comfortable. At this time a rather nice thing happened. I went to see someone in the Admiralty and said that we would very much like to have names for our boats instead of numbers. 'Not a hope, old boy,' he said. 'There's an absolutely rigid rule: anything less than 130 feet long has got to have a number instead of a name and that's that.'

'Does that mean,' I asked innocently, 'that anything over 130 *can* have a name?' 'Yes, I imagine so.' 'Well, the Steam Gunboats are 150 feet long, so what about it!'

And so the Admiralty not only agreed to names but they also allowed us to submit our own suggestions, which was really a very friendly gesture. So each Commanding Officer consulted his crew and we finally agreed on six names. The Admiralty had stipulated that there should be some common factor in the names to bind them together, and I had always wanted to call No. 9 *Grey Goose*, partly because the first boat I ever owned was called that (it was a duck-punt), and partly because grey geese are very romantic birds. So the other five boats became *Grey Wolf*, *Grey Shark*, *Grey Fox*, *Grey Owl* and *Grey Seal*.

We were based at Newhaven and we used to set out at dusk for our sweeps along the enemy convoy route close under the French Coast, returning at dawn. Sometimes we would go four or five nights a week, and sometimes there'd be a spell when nearly every night produced some kind of action. Fierce little battles they were, often inconclusive, fought at ranges of a few hundred yards in the blinding maelstrom of red and green and white tracer bullets, bursting in showers and streams from the ships, ricocheting off the water. All so deadly, but so very beautiful, with the starshells floating lazily down on their parachutes.

The flotilla put up a tremendous record for sea-time during the Normandy invasion the following year, but I can't tell you about that because I had left them by then. Later still, towards the end of the war, they were used for exploding acoustic mines, and when the war was over – well, they'd had their day and it didn't seem that they would last much longer. But then there came a requirement for a hull in which to try out the new type of engine, the gas turbine. The principle of the jet was to be used to drive a boat, and dear old *Grey Goose* was selected. So the wheel had come full circle, and she was back on the measured mile after ten years of loyal service in the forefront of the battle.

From *The Eye of the Wind*

PART III

SPORTSMAN

35

'Gold C' Completed

On 1 July 1957, the clouds at Nympsfield were gangling and loose-limbed like rather bedraggled chickens. To be sure, their tops were in flower occasionally, but their feathery flanks trailed away to a damp and ragged fringe at the bottom. Twice I tried to get up to them, flying alone in the T-42B Sea Eagle glider, but each time I was in weak thermals for little more than twenty minutes. My third launch was at 4.00 p.m. and for half an hour I struggled to keep up. A thermal took me very slowly to 1,500 feet above the site, but an excursion under the most promising vale drew a complete blank. Another weak thermal suggested that it was better to look down for lift than to look up. I drifted back over Stroud, scarcely climbing above my glide-path home, until at 2,000 feet I suddenly noticed a slight improvement, and at 2,800 feet I was among the trailing wisps of a small cloud.

So far I had flown about four times in cloud, and on one proud occasion had climbed over 1,000 feet before using the airbrakes to cure excessive speed. Here, I thought, was an opportunity for some much-needed practice. I wandered into the murk. After a while my circles emerged on one side, so I moved further into the cloud and the lift fell away to nothing. I turned north, flying through patchy cloud with glimpses of the ground. The cloud looked darker ahead – much darker. Almost at once I was in much stronger lift and after a while the altimeter ploughed round in the most purposeful manner. I wondered how long it could last. Six thousand seven hundred feet was my previous highest: Bernie Palfreeman had been to eight-and-a-half: it would be nice to get to nine. A little tuft of water appeared at the front of the canopy by the ventilator, then the canopy began to mist over, and I saw an icicle; then it began to get lighter and much rougher. I pulled down my loosened straps and gripped the stick more firmly. With my other hand I held on tight to the connecting bar of the airbrakes (a safe hand-hold which cannot jerk them open in a bump). The hand of the altimeter wound on past 9,000 feet. I'm not quite certain at what stage the idea of a 'nice little exercise in cloud-flying' changed into an attempt at 'Gold C' height.

A lot of things were happening now. It became very turbulent indeed; the Eagle had no oxygen; and then I couldn't remember how many feet there were in 3,000 metres, but I was too fully occupied to work it out. The speed became erratic, the sailplane was tossed about like an autumn leaf, and then suddenly there was silence and the airspeed indicator swung down to nought: we were in a spin. Now, I felt, was quite a good time to pull out the airbrakes, and when a suitable airspeed had been restored I glanced across to the altimeter – 10,800 feet.

With the airbrakes out I was trying rather unsuccessfully to fly straight and level in a south-westerly direction, and a few seconds later, just twenty minutes after I had gone into it, I burst out of the side of the cloud. It was dazzlingly bright and supremely beautiful: immediately below me was the cauliflower top of one of the foothills of my cloud; and far, far below that again were the straggly wispy strato-cumulus fragments which had been half covering the sky all day. I made a 360° turn to look at the cloud behind me, and only then realised that it was a solitary point – the only big cloud within 20 miles. From below I had had no idea it would be so big. There did not seem to be very much more above, but I was so close to it that perhaps, as with a mountain peak, the summit was hidden by the closest bastion. Most striking of all was the colour – a brilliant golden-yellow in the evening light.

The grandeur of the scene was breathtaking, and I was still under the influence of the glorious relief that the sailplane was under control again. At the same time I was desperately trying to do mental arithmetic and to establish my whereabouts. A thousand metres is 3,281 feet, so what is three thousand metres? Three ones are three, three eights are twenty-four, four and carry two . . . ah! that must be Aston Down, and there's the green reservoir on Minchinhampton Common . . . and carry two, three twos are six and two is eight, three threes are nine – 9,843 plus a thousand for the launch . . . Golly, aren't those clouds beautiful over there under the sun. Well 10,843, but . . . but . . . I only went to 10,800. I've missed my 'Gold C' height by 43 feet! Except of course that I must have lost at least that after the launch and before my first thermal. But supposing the barograph didn't agree with the altimeter? Obviously I had cut it dangerously fine. Incidentally, from that height if I had set off straight away I could have made one glide to Lasham (just over 100 km from Nympsfield) and we wanted the Sea Eagle at Lasham, so that's another opportunity missed. Bother!

By now I was halfway home and down to 7,500 feet, with plenty of sink about. What should I do? If I went back to Nympsfield, maybe Peter Collier could leap in and get *his* 'Gold C' height, but that was rather unlikely as the cloud was now far down wind and no other clouds looked at all promising.

Very gradually it became more and more clear that there was only one sensible thing to do – to turn back, re-enter the cloud, make certain of 'Gold C' height and then head for Lasham. Even after this conclusion became inescapable I found I was still pointing towards Nympsfield. I took a final screw on my courage and swung into a steep 180° turn. The die was cast. I was committed to a further 'short exercise in cloud-flying'.

Five miles to the eastward my cloud still brooded over its own dark shadow lying across the valley at Chalford. It was a bigger cloud now, and it seemed to be higher – a solitary giant more frightening than ever, now that it had come to manhood. Its top was apricot-coloured and crisp, with purple shadows; its foot was murky blue almost merging with the darkness of the Cotswolds below. My first climb had been made in ignorance of the size and quality of the cloud I had been sucked up into, but this time I could see only too well my opponent. I would dearly have liked a really good reason for staying out in the friendly sunshine. I must have been flying in some fairly strong downdraughts, because I only got back to the cloud at about 4,000 feet – a few hundred feet above its base.

I switched on my instruments and plunged into the side, having made a mental note that south would be the quickest way out. At once I found lift and began to circle, but it was not very good, and in searching around to improve it I burst out into the clear air to the north-west. Back inside again there was more lift.

From the beginning the climb was more turbulent than the first. I kept trying to get into the darkest part, in the hope that it would be smoother away from the edge, but it was a vain hope. There was a patter of rain, and later a patter of hail; and there was icing at the front of the canopy as before. Did the hail make it a cumulo-nimbus, I wondered. It was not very loud hail, and there was no lightning. But, cu or cu-nim, I was steaming up at 20 feet per second and bouncing about like a pea in a pod the while.

The altimeter crept past my previous best and up to 11,000, and now the turbulence increased sharply. With no oxygen I had no particular wish to go much higher, but the attempt to straighten up on a southerly course proved disastrous; a few moments later rapid fluctuations of the airspeed persuaded me to pull out the airbrakes. Until then I was still going up, and the altimeter needle now stood at 11,500. I was holding on tight to the airbrake control with one hand, gripping the stick grimly with the other, and occasionally pulling down my shoulder straps which kept working loose.

Still attempting to fly straight and level, I found that the airspeed continued to fluctuate disconcertingly; and then suddenly it increased very sharply so that the indicator read 80 mph. The brakes were out, and at this speed we should have been losing height fast, but instead the altimeter needle was surging round clockwise. Fascinated and seriously worried, I watched it go up 700 feet in about thirty seconds before, with a frightful bump, we flew into a violent downdraught. A minute later the airspeed shot up again, and once more with full brakes out and 80 mph on the clock we climbed 700 feet; again there was a violent jerk and we were going down and a few seconds later we were out in the blessed evening sunshine.

The panic was over. I tried to shut the brakes but they were frozen open. Ahead two great walls of cloud were closing together and I aimed at the gap. The Sea Eagle just squeezed through and I could almost hear the clang as they met behind me. I tried the brakes again and they closed halfway but no more. And so my hard-won height fell away in a miserable glide.

Nevertheless, I headed for Lasham. The total distance from Nympsfield was 65 miles, but already my cloud had taken me 15 miles to the eastward, so that I had barely 55 miles to glide. From 11,000 feet in still air, with the airbrakes in, Lasham should have been in the bag with a little to spare. But with the airbrakes out and the sky full of 'sink', it was quite another kettle of fish.

I flew out over the Cerney gravel pits half hidden by wispy cloud and headed for Swindon. Every minute or so I tried the airbrakes again and managed to get them a little more shut, but I was down to 7,000 before they finally clicked home. To get to Lasham now I must obviously find some more lift. Over Marlborough was another big cloud, and beyond this and to the east were a couple more. I headed out west of my course to the Marlborough cloud. I was a few hundred feet below cloud base when I got to it . . . and found nothing. A bonfire was burning at the edge of Savernake forest and the smoke came up towards my cloud. There was even a tuft of slightly lighter cloud at the point where the bonfire's hot air appeared to enter the darkness of the decaying giant. Round and round I went, muttering the familiar sailplane pilot's *cri de coeur* – 'There must be something here somewhere.' But the best I could find barely halted my descent. At a quarter past six in the evening it was perhaps only to be expected.

I headed on, working desperately at my gliding-angle graph, and soon saw that Lasham was just beyond my grasp. Clearly there would be no more lift, so I had the choice of landing in a field a few miles short or turning off to the nearest aerodrome. Andover seemed to be within reach and from there I could be aero-towed to Lasham; so to Andover I went – 48 miles from Nympsfield. The long glide was not without its anxieties, for the sunlit town far out on the horizon seemed an impossible distance, but when I remembered that I had 500 feet more than I had thought because of the differences in airfield height, I was able to sit back and enjoy the beauty of the countryside in the orange evening light.

I landed at a quarter to seven and not long afterwards my barograph chart was being signed by Andy Gough.

—

The quality of sailplanes is now such that 'Gold C' distance flights are two a penny, but as they are also tremendous fun, why should we not make them three a penny? Not for nothing was I a Boy Scout, and I am more than ever convinced that most of the success of a 300-km cross-country flight depends on the plans you have laid beforehand. You must be absolutely ready, absolutely early enough, on the day of days.

Perhaps the most important thing is to have a clear conception of where you are going. Or so at least I found it on 12 April 1958. St Just aerodrome is almost at Land's End, and 298.8 km from Nympsfield, so a dog-leg is required, but as the turning point does not have to be declared I had drawn three arcs on my map. At each end was an arc of 80 km, one from the point of departure and one from the goal. Between the two arcs was another obtained by joining Nympsfield and St Just with a string representing 300 km, which of course was slightly slack, and by pulling it out to one side and sliding a pencil along the inside of it from one 80-km arc to the other, I marked this out on the south side of the direct line and could then say that any recognisable point outside the three arcs would be an adequate turning-point for the 300-km flight. I also knew that the far end of the Cornish Peninsula was proverbially devoid of thermals, especially in the late afternoon. And so I decided to go as far outside my arc as possible, so that even if the tip of Cornwall let me down, and the goal could not be reached, there was some chance of getting the distance.

On the morning of the great day, which was a Saturday, there was a conspicuous lack of activity at Nympsfield although the sky was already full of cloud. Peter Collier was fussing over the Club Skylark II and Derek Stowe (who had made the Land's End flight last summer but whose photographs were questionably adequate) was preparing the Club Olympia. There was a winch driver but not enough people to hold wing-tips on the way to the launch point. I towed out the Skylark behind my car and John Hahn came with me to the individual hangar which houses the Sea Eagle. When we got back to the launch point, with the Eagle behind, all was in readiness for the first launch of the day, and at about 10.20 Peter began to circle off the top launch. By the time the cable had been retrieved he was already at cloud base and fast disappearing down wind.

My own launch was at 10.37 and I fumbled the dregs of Peter's thermal. I was down to 400 feet before I found another and I, too, was away. As I circled up towards cloud base I watched the blue Olympia arriving at the launch point and finally taking off. Here were the three of us bound for the same goal. I think it was this 'competition atmosphere' that frightened me most. After all, Derek Stowe had been there once already and Peter

Collier had taught me to fly a glider. Somehow I must contrive not to be the first of the three to land.

The 12th of April was not, in my view, a 'day of days', although of course it was a very good day. As I meandered slowly down the West Country I never climbed more rapidly than 400 feet per minute, but with 20 miles an hour of wind behind me and an early start, I reckoned I had only one job to do and that was to stay airborne. Not for me the wild down-wind dash between thermals; not for me the 70 mph average. All I needed to do was float lazily along on the wind, dallying with every flutter of lift that I came upon.

I had some qualms when crossing the dreaded Taunton Vale. I performed some involuntary aerobatics while using both hands to photograph my turning-points first at Weston Zoyland and then at Culmhead which was a bit farther outside the arc. Just past Exeter I spotted a red Olympia circling three miles astern of me, but I have not yet discovered who it was. Over Dartmoor the only respectable cloud was in the very centre and rapidly decaying, but the dregs of its lift wafted me to the green fields beyond.

By the time I climbed in a glorious blue thermal which steamed up amongst the wreckage of a decaying cloud over the china clay mines near St Austell, I reckoned that I had a good chance to make my goal. But in cross-country soaring everything seems to be going so well until the moment when suddenly it isn't. At 5,300 feet I turned westwards from the china clay mines into a dead blue sky. Hastily I looked at my graph and worked out that at 20 mph of following wind and 40 miles to go the outcome lay delicately in the balance. I flew as accurately as I could at 43 knots. And as I flew my anxiety mounted. For whatever happened now I was in for one of those horrible marginal 'glide-outs' which left one making decisions much too near the ground. At first it all seemed to be going rather well, but then surprisingly I began to sink more rapidly. This I hoped might mean a new brew of thermals, and indeed tantalisingly a tiny puff of cumulus appeared over the cliffs at the very tip of the peninsula. But it was not to be.

At last it was evident that I could not conceivably work my way round the high ground to St Just airport. And indeed in my calculations I had forgotten to add the airfield height of 300 feet of St Just itself. So I turned left down wind into the low country stretching south-westwards from Penzance. I was getting very near the ground, keeping one field up my sleeve and assessing another ahead. They were all horrifying small. At last it was clear that I could not top the next ridge and I must turn back immediately. I swung round not more than fifty feet above a tiny field into a wind which was gloriously stronger than I had expected. The Eagle hung delightfully in it, sat gently down and came to rest in 48 yards, 20 yards short of the wall and 305 km from Nympsfield. It was five hours to the minute since I had taken off.

From *Sailplane and Gliding*

36

Straight and Level, Please

Most of us glide because it stimulates us. What we mean is that it frightens us. The question is: 'How much?' If it frightens us too much we no longer enjoy it; if it does not frighten us enough we become bored. Somewhere between the two lies the special appeal of our sport, and it leaves a fairly wide scope for individual variation.

On this general basis I myself have always been a member of the Straight and Level Club. Aerobatics are only enjoyable to me in their mildest forms. Chandelles are all right, loops are my limit, and anything which disturbs the dust from the cockpit floor is well beyond it. Thus when I arrived at the World Gliding Championships at Leszno in Poland I was astonished to see a two-seater Bocian glider circling at 500 feet over the middle of the aerodrome *upside-down*. It was not until the following day that Colonel Benson and Teddy Proll told me that they had both been indulging in this doubtful entertainment and that if I wanted to fly during my short four days at Leszno they could arrange it. At this point I thought I made myself clear that, although there was nothing I should like to do more than to sample the Polish thermals, there was nothing I wished to do less than to fly upside-down. An hour or so later Teddy Proll approached me excitedly and said he had arranged it all and that in a few moments I could have a flight in the two-seater Bocian.

Now, this machine was not entirely new to me because, a

week before, I had flown in one at Helsinki and had found it, although not to my way of thinking quite so attractive as an Eagle, nevertheless a perfectly gentlemanly aircraft of high performance and considerable comfort. I therefore looked forward to my flight with lively anticipation and soon afterwards, at the launch point, I was introduced to the charming young man who was to fly with me. He spoke a very few words of English and a very few words of German, and to make quite sure I explained to him at some length in both languages that the limit of my ambition was perhaps an hour of comfortable thermal soaring, after which I hoped to return equally comfortably to earth. He seemed an intelligent young man and appeared to understand me perfectly. There was some delay because on the previous landing the airbrakes had become inexplicably jammed. When the seats were removed a screwdriver was found wedged behind the airbrake control. Apparently during inverted flight it had slipped out of a pocket. At least this particular hazard would not come my way. No inverted flight for me.

We were launched by aero-tow with a 60-foot tow rope of suitable thickness for towing a motor car. I found that the climb to 500 metres required all my attention, for the tug – a low-wing monoplane – was terribly close in front of us. To begin with, when I flew in the standard position, with the tug's wings just below the horizon, I was looking down into the cockpit and could almost read his instruments. I was shortly corrected by my young colleague and told that I must keep all of the tug above the horizon. This I did and was surprised not to fall into his slip-stream, but it turned out to be quite a comfortable position. Nevertheless the short tow-rope still required a high degree of concentration.

At 500 metres we released in an indifferent thermal and I started to turn. But my companion shouted, 'No, no! Not yet!' and seized the controls. We flew farther into the thermal, which had admittedly improved, and was of such enormous dimensions that the straight period did not carry us through to the other side as it would have done in England. As we began to gain height I found that my companion was a confirmed 'pudding-stirrer'. The control column was in constant motion and as a result (so it seemed) we had quite a rough ride. After a while the thermal grew weaker and I asked if I could fly again. Here I was lucky, for I decided to move over to another thermal underneath a cloud which was just forming, and this was so much stronger that we roared up at about 500 feet a minute.

From 5,000 feet we headed back towards the aerodrome. 'Now aerobatics!' said my friend. Could it be that he had misunderstood me? Well, there was no harm in the simple ones. I performed a couple of fairly mild chandelles followed by a loop. 'Is very nice,' said my friend, 'now I show you.' He took over the controls and in a second we were flying upside-down.

A number of unfortunate circumstances dominated the next few moments. First, I had not taken the elementary precaution of tightening the lower pair of straps. Tightly though my shoulders were held, my midriff was only loosely supported. In order to offset this disadvantage I had found a convenient hand grip for my left hand under the seat. Everything seemed under control although I was hanging rather far away from the seat itself. A few moments later my companion began a turn. At this point the seat, unaccustomed to an 'upward' pull of one and a half times my weight, gave way with a splintering crash. This was the signal for a loud guffaw from my companion. As a child I wonder whether you can remember lying belly downwards in your bath. I was in just such a position, only the bath was a Perspex canopy. It was at this stage, to my undying shame, that I could no longer withhold a stifled cry for mercy. With a flip back of the stick we were flying the right way up again.

Hastily I tightened the straps, for clearly we had not seen the end of this business. In a few seconds we were involved in two consecutive slow rolls. But with the straps tighter I felt slightly more secure, although my toes were curled around something – perhaps the variometer bottle – which was quite certainly not designed to take the strain now bearing upon it. The next thing was a half-loop to inverted-flight, and at that precise moment we hit the edges of a thermal. 'Ah-ha,' said my companion, and we began to circle upside down. It was only when he started to tighten the turns and we had already gone up 50 metres that I allowed a further expression of dismay to escape my lips. This time, in what I hoped was a firm voice, I followed it with, 'right way up now, please'. A few seconds later a more normal world was restored to me.

We were still regrettably high. 'Now Immelmann,' said my companion, and in quick succession we performed two half-loops, rolling off the top. After what I had already suffered, these were, it must be admitted, comparatively mild and I even made so bold as to try one myself, but it was executed at too slow a speed, and my companion demonstrated with two more.

As we now approached the lower limit of what in Britain would be regarded as aerobatic height I began to breathe again, but my relief was premature. First came two rather charming little flick stall turns, a manoeuvre which only leaves you on the straps for a second or so.

We were now down to 500 feet. Down went the nose yet again. 'What now?' I thought. At the edge of the aerodrome our Bocian half-rolled on to its back and we made a run across the whole width of the field upside-down. Oh dear, oh dear! But surely there could not be much more. We half-rolled out, went up into a chandelle, round, out brakes and a spot landing which trickled us up to the hangar door. An enthusiastic Teddy Proll rushed up to take a snapshot and to ask the inevitable question: 'Did you enjoy it?'

From *Sailplane and Gliding*

37

To Make the Spirit Soar

Last Tuesday a big cumulus cloud took me up in my sailplane to 8,000 feet and then, getting bumpy and turbulent near its top, it persuaded me to leave. I turned on to a westerly course and less than a minute later burst into the sunshine. The scene into which I emerged was so supremely beautiful that I exclaimed aloud. Huge cloud masses surrounded me, golden yellow and deep blue grey in the evening sunshine. Away to the north was a vignette of a distant cloud giant framed by the nearby bastions. Ahead, gleaming up through wispy layer-cloud, was the Bristol Channel and the Severn Estuary, lying like a great silver bar across the greenness of England. Far, far below was my home with its surrounding duck ponds and the ribbon of the Sharpness-Gloucester canal.

It was a scene to make the spirit soar, a glorious and memorable experience. It was much better than anything I thought could happen to me when I first took up the sport of gliding two years ago. There is a special quality in those sports where a man may pit his skill not merely against his fellow man or some other animal but against the elements. Adventure comes in when the opponent is a wind-swept mountain or a gale at sea, or a great black thunder cloud. I believe that climbers dislike the suggestion that they have 'conquered' a peak, and perhaps ocean-racers do not consciously regard themselves as competing against a storm and winning if they survive unsunk, although the meteorologists' device of giving names to hurricanes perhaps strengthens the notion. Certainly to the glider pilot there is no inanimate object with such a living personality as a cumulo-nimbus cloud. Its brooding darkness, its lightning and its thunder, its unimaginable power which can suck a glider up to the height of Mount Everest in half an hour, all combine to make the pilot view the creature with awe, if not with dread: and yet it contains in the most generous measure the very thing he is looking for – lift. Against all his instincts he is attracted to its black belly like a moth to a candle.

From *The Sunday Times*

38

Prince of Wales's Cup and Olympic Games

In the race for the Prince of Wales's Cup at Falmouth in 1934 third place was taken by the fourteen-foot dinghy *Whisper* belonging to David Drew. She was probably no faster than *Eastlight* but she was the prettiest of all the boats in the fleet; her timber had been very carefully selected and the most glorious grain continued across her sides from plank to plank, so that she appeared to have been carved from the whole trunk of a tree. Her colour was that beautiful purplish-chestnut of some kinds of mahogany whereas *Eastlight*'s colour was yellower, and seemed less mellow. *Whisper* was so lovely, and incidentally had done so well in the light airs, that I decided to sell *Eastlight* and buy her. The price, as I recall, was £130. *Eastlight*'s number had been 318, *Whisper* was 324; although she was six boats later, the vintage was the same, and her performance was almost identical.

In 1935 the Prince of Wales's Cup was to be raced for in Osborne Bay in the Solent and again Nicky Cooke was to be my crew. We had tuned up *Whisper* to a high degree, and had done well in the earlier part of the season. Once more we were full of confidence.

On the day before the big race we were out in front, doing very well in a fresh breeze, when one of the rigging wires of the mast broke and the mast itself bent alarmingly. I let the sheet go just in time and the mast was saved. The wire which had parted was the bottom diamond wire on one side, part of a stressed structure of single-strand wires on spreaders of aluminium tube which kept the mast straight. Congratulating ourselves that it had broken the day *before* the big race and not during the race itself, we got the boat back to Cowes, whipped out the mast and took it up to Uffa Fox's yard where the boat and its spars had originally been built. A new diamond wire was fitted in the place of the broken one, and all seemed to be ready for the race the next day.

It was another day of strong winds, which suited us perfectly. Nicky and I went off on the first beat without a care in the world, and before long it was clear that we were in the lead again, just as we had been on the day before. As we approached the weather mark I suddenly saw that the mast was bending as badly as yesterday. The wire had not broken, but it seemed to have stretched, or else the wire knot had slipped at its attachment. We went about quickly on to the other tack, and found that the mast stood up quite straight. It only needed one more short starboard tack to get ourselves round the weather mark, still first by nine seconds from Stewart Morris. On the broad reach Nicky did his best to tighten up the slackened diamond wire, but it had stretched too much and the rigging screw would not take up enough to get it taut again. Round the leeward mark we came on to a port tack where the mast stood up quite well. We plugged on, but in due course we had to tack and as soon as the sails filled away on the starboard the mast began to bend again. We gilled along gently not daring to fill the sails, and then, as we sailed only at half power, we found ourselves overtaken by Willie Hicks from North Norfolk, who had been going very well on the wind. He crossed us once and we were faced with a major decision. There were three and a half more rounds to the race. If we sailed on at half power we should drop down and down. There was just a chance that if we drove her hard, the mast would bend until the slack diamond wire was tight, without actually breaking. We decided to haul in the sheet, drive her as fast as we could, and hope for the best. As we hauled in, the boat surged forward again, but the mast bent like a bow; it was clear that it would not last long. A moment later we came over the crest of a big wave and with a splintering crack it broke. The rest of the fleet sailed past, and that was that.

I was bitterly angry. I felt we had been let down by bad workmanship, though no doubt the chap who had fitted the diamond wire had done his best. I found it especially hard to be a 'good loser', even though I knew I would lose more by being a bad one. Our misfortune in the Prince of Wales's Cup of 1935 was doubtless very good for the character.

Steward was the winner, John Winter was second and Willie Hicks third. It was small consolation that *Whisper*, with a new mast, won the race on the following day in identical conditions over the same course. It was *not* for the Prince of Wales's Cup.

The venue of the championship changed annually, but for weekend racing the keenest competition in the fourteen-footers was in Chichester Harbour where a strong fleet, including most of the best helmsmen in the class, was assembled at Itchenor.

One day after a race as we were hauling our boats out of the water at the end of the club jetty, I noticed a small sailing dinghy lying a few yards offshore. At her tiller was a boy wearing a pair of blue bathing pants. He was mahogany brown from sunburn, his hair bleached to the colour of straw; and his blue eyes were almost popping out of his head with excitement at seeing, at close quarters, the beautiful International fourteen-foot dinghies which he had watched only at a distance during their races. I did not know that he had sailed round from Emsworth especially for the purpose, but I could see how much he wanted to look at them more closely, so I called across to him:

'You can come up the slipway if you like and look at mine.' Stewart looked reproachfully at me.

'He can't possibly come up unless he puts a shirt on,' he said. 'The older members wouldn't like it.'

So I called across, 'Have you got a shirt?' And the boy said

'Yes' and put it on. After he had examined the dinghies for some time I could see that he longed for a sail in one.

'All right,' said I. 'Next weekend, if you come over, I'll take you out in *Whisper*; by the way, what's your name?'

'Charles Curry.'

'Right, see you next Saturday.'

With a fresh breeze Charles and I took *Whisper* down the harbour a week later and planed backwards and forwards at speed in the most exhilarating clouds of spray. After about an hour when we were thinking of coming back an upper diamond wire suddenly parted and the mast fell over the side again. We drifted in to the shore, waded through the mud to the salt-marsh and walked disconsolately back to the club to get a motor boat to go down and tow the dinghy home. That was Charles Curry's first introduction to the fourteen-footers.

Early in the summer of 1936 I decided to enter for the trials for the single-handed sailing in the Olympic Games. The event was to be sailed in the German Olympic Monotype, or Olympia-jolle. Three of these boats had been obtained by F. G. ('Tiny') Mitchell, Commodore of the Royal Corinthian Yacht Club and brief eliminating trials were held at Burnham-on-Crouch. By some extraordinary chance I won these trials. I still do not quite know how it happened, but I suppose you cannot have bad luck all the time. Stewart was second and became my spare man for the Olympic Regatta at Kiel. At the time the importance not only of the trials, but of the yachting events in the Games themselves was surprisingly low. It may seem extraordinary now that neither Stewart nor I (nor indeed any of the fourteen-foot dinghy sailors) rated the Olympic Yachting as in any way comparable in importance with the Prince of Wales's Cup. In this judgement we were probably right. The general standard of skill shown by competitors in the Prince of Wales's Cup was almost certainly higher than in the single-handed Olympic contest.

Chronologically the Prince of Wales's Cup came first. It was to be held at Hunter's Quay on the Clyde and we had been told that light weather might be expected – indeed the Clyde was famed for it. So I agreed to sell *Whisper* to Michael Bratby and from Uffa Fox I ordered a new boat, to be called *Daybreak*; she was to have the minimum beam in order to reduce the wetted surface, and would therefore be less good for planing. This was the gamble I had to take.

Immediately before the Prince of Wales's Cup races we were to sail a series of team races against the United States. My crew, both for the team races and for the Prince of Wales's Cup itself, was once more Beecher Moore.

As well as being narrower than most of the other boats of her class *Daybreak* had a secret weapon: we had decided to try a wooden centreboard. We had no idea whether this would be effective, but until we knew, we were determined that it should remain a secret.

At this time the centreboards of most of the dinghies were made of a single plate of phosphor-bronze – a dull, brass-like metal – weighing something like a hundred pounds; the wooden centreboard weighed ten pounds. The boat would be enormously lighter. But should we miss the stability of the lower centre of gravity? In order to preserve our secret we painted the light board with brass paint, and whenever we carried it, we put on a great pantomime of weight lifting.

Beecher believed in preserving the right state of mind in his helmsman. More recently he has compared the relationship of crew to helmsman with that of a farmer to his prize pig. I was treated to every kind of luxury in order to put me at my ease, and on the Wednesday – the day before the Cup – I was not allowed to race, but spent a most delightful day in the hills to the north of the Clyde with Beecher and his wife, trying to think about something other than sailing.

But on the Thursday, it blew a gale, so hard that the Committee decided to postpone the race. Although a gale of wind was not likely to suit our *Daybreak*, Beecher and I were fairly heavy and competent when it was blowing hard, so we felt annoyed and frustrated when the race was postponed. On the following day – the Friday – the wind was as strong as ever and there was another postponement. Again we made our protest to the Committee. The only other person who minded was Robert Hichens, a Falmouth solicitor who had a home-made boat called *Venture*.

By the Saturday morning, the last day of the week, the gale had abated a little though not very much. At about 9.00 a.m. Robert and I, having protested, were asked by the Committee at Uffa's suggestion, to sail out as guinea-pigs. We were to test the course and see if the weather was fit for the race to be held. Out we planed, from the shelter of the town and the hillside into a strong wind and a very big sea. We could almost feel the eyes watching us from the shore. With our mainsails reefed we were never overpowered. When we got back Robert and I agreed on our report.

'Yes, it is just fit to sail today, but it would certainly not have been on Thursday or Friday and we both unreservedly withdraw our protests.'

The start was postponed till one o'clock in the hope that the wind would moderate, but there was little change. It blew and blew and rained and rained.

We made a good start and got to the weather mark thirty-one seconds ahead of the rest of the fleet. This was a dramatic and impressive lead. To windward it seemed that *Daybreak* was unbeatable, but there were two more legs of the triangle and three more rounds after that. How would she compare at planing for which she had not really been designed? I remembered the saying, which was still true, that in no previous race for the Prince of Wales's Cup had the boat which led at the first mark in the first round ever been the winner.

The old gang was out in front again. John Winter rounded second, with Stewart only seven seconds behind him in a field of thirty-one. Then the planing began. *Daybreak* planed very fast, but *Lightning* planed much faster. She did the second leg of the course in one minute six seconds less than *Daybreak*, planing through our lee in a great sheet of spray. It was depressing, yet no more than we expected; not only was *Lightning* designed as a planing hull, but John was far better than me at starting his boat off, and keeping her planing, even if I could usually beat him to windward.

We stayed second for another round, then Stewart who had been biding his time, came up on the reach and planed past us just as John had done. John led for the third round but I had the feeling that Stewart was awaiting his chance to pounce. Not only was he planing as fast as John but he was going very fast to windward too. Then John struck a bad patch; he had shipped a lot of water in his boat and his crew was forced to bail on the beat. While that was happening Stewart took the lead and won the race for the fourth time. John finished second for the second time and *Daybreak* came third to win my first replica for me. Phyllis Richardson was fourth and Robert Hichens sixth. Charles Curry had been sailing *Whisper* at the invitation of Michael Bratby who crewed him, and they were doing quite well until they capsized at the gybe in the second round.

So it had not been 'third time lucky'. Perhaps in *Eastlight* or *Whisper* it would have been a closer thing, but the truth was that two helmsmen in fourteen-footers were still substantially better than me; although I could often beat them in lesser races, and had managed twice now to get myself out in front at the beginning of the Big Race, they were in front of me at the time it mattered most – when crossing the finishing line.

The excursion of the fourteen-footers to the Clyde was memorable for a delightful story to which Michael Bratby bore witness. The two postponements of the big race had greatly disorganised the plans of those who wanted to return south at the end of the week. After finishing fourth in what must have been the most strenuous race of her life Aunt Phyllis was eager to catch the night train. Leaving her husband to pack up the dinghy, she was driven to Glasgow by Michael. They arrived at the barrier just as the guard was waving his green flag. The wheels had already begun to move when Aunt Phyllis, drawn up to her full six feet, shouted in an imperious voice,

'Stop the train – I'm Mrs Henry Richardson.'

The name could have meant nothing to the guard but the tone of voice was not to be denied; he blew a shrill blast on his whistle and waved his *red* flag. The signal was repeated farther up the platform and the train stopped.

—

Although we had placed the Olympic Games far below the Prince of Wales's Cup in importance in the season's racing, my failure to win the Big Race became an additional spur in the regatta at Kiel.

Apart from the initial trials my only experience of the boat I was to sail had been a week on the Buiten Y, a part of the Zuider Zee near Amsterdam, in early June. Dutch, German and Norwegian helmsmen were racing and Stewart Morris and I were both lent boats for the regatta. In spite of the distraction of low-flying spoonbills flighting out from the Naarder Meer, I had managed to win the points cup for the week, with two firsts and two seconds, and Stewart came fourth with one first and three thirds. Neither of us had liked the Olympic Monotype as a boat, but of the two Stewart seemed to like it the less. In spite of the Clyde, I was determined not to allow any inferiority complex to prevail. After all, I had won at Amsterdam.

In Germany, Stewart was much more than my spare man; he was my team manager and my coach, and looked after me and my boat with meticulous care.

The series began on 5 August 1936, and in the first race it blew hard. In that weather I had the measure of all the other competitors, many of whom had very little experience either of strong winds or of the Monotype. The more it blew, the more I felt at home, though the boat, of course, was fairly new to me too. She was sixteen feet long, rather beamy, and was 'cat-rigged' which means that she had only a mainsail – no jib. The rather heavy mast was stepped very near to the bows, and the boat became extremely difficult to control off the wind in a blow. The tiller was a huge wishbone affair, which was the standard continental method of getting over the difficulty of a helmsman who has to steer while leaning far out of the boat. Its weight and inertia alone greatly reduced the delicacy of the control. By comparison with a fourteen-footer the Olympia-jolle seemed a very blunt instrument.

The German word *Jolle* is etymologically related to the English 'jollyboat' and the English 'yawl', but in yachting circles a *jolle* and a yawl are very different vessels: a *jolle* is a small dinghy; a yawl is a yacht with mainmast and mizzen, and is usually of several tons register. The soft *j* of *jolle*, however, led to an ignorant translation of Olympia-jolle into Olympic Yawl, by which name the class became known throughout most of the British Press.

On that first windy day I went away to win quite comfortably from the German Willi Krogmann, and that night we dried the sail in the cruiser *Neptune* which was lying in the harbour at Kiel. But in the next few days the wind fell away and my performance was gradually outshone by the Dutch helmsman Dan Kaghelland.

By 11 August – the day of the last race in the series – Dan was so far ahead on points that he had won the gold medal. He had no need even to start, though in fact he did so. Will Krogmann and I had an exactly equal number of points, but again we were so far ahead of the Brazilian who was lying fourth that he could not possibly overtake us. So far as the gold, silver and bronze medals were concerned, the last race had only to decide whether Willi or I should have the silver one.

At the start we manoeuvred with eyes only for each other and Willi got the better of it and went away ahead of me. Then followed a spirited race in which we ignored the other twenty-five competitors entirely. By the second round when we lay respectively thirteenth and fourteenth, I was coming up on Willi fast. The wind had freshened and it was quite clear that in a short while I should overtake him. He tacked to cover me and we sailed along side by side going to windward on a starboard tack. At the outset of this tack his boat lay about fifty yards to weather of mine, and he was therefore in effect still about fifty yards ahead. In the fresher breeze we both lay out to the very limit of our strength. This was it. If he could hold me back now, the silver medal was his, if I could get past him it was mine. I was gaining on him by sailing an infinitesimal amount closer to the wind. The two boats remained level, almost stem to stem, but their courses were gradually converging; I was shortening that fifty-yard lead. As I sat out on the starboard gunwale Willi was directly behind me. I did not need to look round at him; I could hear his bow wave, and by its direction I could tell that we were both sailing at the same speed, but by sailing closer to the wind I was still closing the gap.

Of a sudden the wind changed its direction very slightly and both of us were able to point a little higher than we had been pointing. This minor change in our positions relative to the wind's eye, meant that the wind shadow from Willi's sail now fell across my sail. There was nothing for it but to tack immediately. I scarcely looked behind and flung the boat round. To my consternation I felt a little tap as she went round and realised that I had nicked the stern of Willi's boat. This of course I had no right to do; it was a collision and the fault was most certainly mine. Although I knew I had been gaining on the German boat, I could not believe I had reduced that fifty-yard lead to a boat's length in so short a time. But there was a second reason. In the quick glance behind me I seemed to have room to tack astern of him, but I was thinking in terms of a fourteen-foot rather than a sixteen-foot boat. My familiar fourteen-footer would have cleared him easily but the Olympic Monotype did not.

The damage was done; there was, as I saw it, no alternative

but to retire immediately. I dropped the mainsheet, bore off down the wind and sailed out of the race. It was a very bitter moment and small solace that it was regarded as the correct thing to do.

One way and another 1936 had been a full year: apart from winning a replica of the Prince of Wales's Cup and an Olympic bronze medal – in spite of disappointments, I was very proud of both – my book *Morning Flight* had been published in its first 'Ordinary Edition', I had had an exhibition of pictures at Ackermann's and had illustrated my stepfather's book *A Bird in the Bush* which was published that autumn. It was very well received and reviewed, but I was not at all satisfied with my drawings. I knew that they were not in the same class of excellence as the text. A few of the designs were fine and bold but with too many of the small birds I was groping for a likeness. For so exquisite a book the illustrations should have been perfect and I was sad that they were not. It was a poetical book and by chance Bill was President of the Poetry Society, as well as Minister of Health, when he wrote it.

Birds and sailing had met on the common ground of East Anglia for so many summers now that my family decided we should have a permanent country home in that part of the world. So we came to live in holiday times at Fritton Hithe, an unobtrusively Victorian house thatched with Norfolk reed, which overlooked Fritton Lake, near Great Yarmouth. In front of the house, which rambled about, mostly at ground-floor level, was a wide lawn leading down to the reed-fringed lake side and a small jetty from which we could bathe and fish and sail, though the lake was so sheltered by the surrounding woods that there was seldom a steady breeze. This did not prevent us from sailing a small fleet of eight-foot Corinthian Otter 'prams', with tiny balanced lugs. We had two of them, and our neighbour across the lake, General Sir Thomas Jackson, had two more for his family of two sons and a daughter. It was in these that my brother Wayland learned to sail.

The care of the 'prams' was my stepfather's special province, and I think it was this that led him to study knots and splices. It was astonishing that he could make the most complicated turk's heads, and could sail those tiny, tipply boats with only one arm.

He had lately retired from politics and had become Lord Kennet of the Dene, his title taken from the river and the little valley at Lockeridge. At one time it had seemed that he might go from the Ministry of Health to the Exchequer, but Baldwin withdrew his offer, and there was then no alternative but the House of Lords. In face of what was a palpable breach of faith my stepfather showed no signs of disappointment or bitterness, though my mother was not so charitable about the Prime Minister's change of mind, and particularly about the way it was done.

I never admired Bill more than during the critical time when his political career was in the balance. Had he become Chancellor of the Exchequer then, he could conceivably have been the next Prime Minister, but at home it was impossible to tell that any such things were in his mind. Basically, perhaps, he was not ruthless enough for the political pinnacle.

At the east end of Fritton Lake were three decoy pipes still operated by our landlord, Lord Somerleyton, on a minor commercial scale, and in winter there were frequently large flocks of ducks in the hidden corner of the lake. In summer great crested grebes displayed at the foot of the lawn.

My mother took great delight in the garden at Fritton. There was a brilliant herbaceous border, on which, in season, great numbers of comma butterflies were to be seen, and once a large tortoiseshell. At the back of the house were two small greenhouses – in one of them a splendid vine which kept us in grapes for half the summer it seemed; in the other exotic plants like *Pamianthe peruviana*, *Gloriosa rothschildii* and a few orchids. It was the first time we had ever had greenhouses and my mother was very proud of the beautiful things she grew in them.

Fritton was a part of my life for about twelve years – a home for a family in peace and war. Distinguished musicians played on its two grand pianos, artists, writers, and politicians came for the weekend or the holidays and played parlour games (J. B. Priestley as a St Bernard dog was never to be forgotten), but in particular the house was full of young people, contemporaries of my brother and me. This was especially so during the regattas in the summer.

Sea Week at Lowestoft in 1937 was more than usually important because the Prince of Wales's Cup was to be held there again. For this, my fourth attempt, I had commissioned Uffa to build me a new dinghy, called *Thunder.* (As owner of *Lightning* John Winter claimed that his must always come before the thunder, but I was not to be deterred.) I had persuaded Charles Curry to be my crew.

In the previous year *Daybreak* had been designed for the light airs on the Clyde and we had raced in a gale. The weather at Lowestoft was anyone's guess, but as my best chances seemed to be in a strong wind I had decided to gamble again – to have a dinghy with a planing hull. Already I knew that sailing to windward in a blow there were few who could beat me. If I could plane fast off the wind I should have a good all-round chance in fresh breezes; but by the same token if we had the light weather we should have had in Scotland I would be at a similar disadvantage in reverse.

Thunder was the most beautifully finished sailing boat I had ever seen. This perfection of functional efficiency was reflected in the boat's appearance and was part of its particular appeal. Not only were the fourteen-footers thoroughbreds, but they looked like thoroughbreds. They were built with the precision and artistry of a violin; Uffa had set new standards of workmanship in boat building, and to own one of his fourteen-footers in the 1930s was to own the most perfect little boat in the world.

Although Stephen Potter had not yet invented the phrases, there was plenty of gamesmanship and even lifemanship to be practised in dinghy sailing. For several years a secret preparation had been applied to the bottoms of certain boats on the morning of the Big Race. It was called 'Elgo', was supposed I to make the boat 'go like 'ell', and consisted of the whites of half a dozen eggs applied with a rag and allowed to dry. Almost certainly all traces of it were washed off by the time the boat had been in the water five minutes, but for many years it persisted as a supposed secret weapon available only to a privileged few. Before the 1937 race we decided that a new gambit was required. I let slip that I had been secretly examining the cost of an ultra-light aluminium mast for *Thunder.* On the day before the race we rubbed down the existing wooden mast and painted it with aluminium paint. Until that time all dinghies had had varnished masts, but the silver colour looked in many ways more attractive and racey than the yellow varnished silver spruce. On the morning of the race we stepped the old mast in its new colours and all round the dinghy park tongues began to wag. In the course of the morning one or two people came up, tapped the mast gently and immediately realised that it was only wood. But among the fleet of forty-six dinghies, few had time to do this and there were many who believed that we were sailing with the expensive advantage of a lighter mast. This went only a little way to offset our misgivings about the weather. The day had dawned absolutely calm and there was little hope of more than a zephyr all day. I still owned *Daybreak* but had chartered her to David Pollock, and now it was *Daybreak*'s weather. There was nothing I could do about it but watch David take the lead almost from the start. In the very light airs the windward leg of the race had to be sailed close in to the beach because of the strong south-going tide. The whole fleet stood in towards the shore and for a mile short-tacked among the groynes and holiday bathers, and past the Claremont Pier with its hazard of fishing lines.

We were away close behind David and stayed close upon his tail. I felt confident that we should find an opportunity to pounce on him at a later stage in the race but to begin with he was offering us no chances. We were a good deal more concerned that John Winter and Stewart Morris were not very far behind us. Early in the third round *Thunder* went into the lead and as we tacked up the shore we had a difficult decision to make. Each tack was very short – perhaps forty or fifty yards out into the tide, then 'about', and forty or fifty yards back again towards the beach, the crew with his hand on the centreboard rope ready to pull it up the instant it touched. The wooden breakwaters jutted out at frequent intervals, making helpful tidal eddies, but complicating the defensive tactics required of a leading boat. I must try all the time to 'cover' the next astern – to be on the same tack as her, and whenever possible to get between her and the eye of the wind, so as to reduce her motive power. Every time John tacked I too must tack so as to be similarly affected by any small change in the direction of the wind. Being in front my tactics were to maintain the *status quo* and take no chances. The boat behind on the other hand was always trying to get on to the opposite tack.

In this case, however, John was covering Stewart so that in looking after one I was looking after them both. But my problem was suddenly complicated when John allowed Stewart to break tacks and go off on his own. My two most important rivals were now on opposite tacks. Which of them should I cover and make sure of keeping behind me? Stewart was four times winner of the Cup and was sailing *Alarm*, in which he had won the race in the two previous years. John had won the race once at Falmouth and was still sailing *Lightning*, the boat in which he had then triumphed. Since then he had twice been second in the same boat. *Lightning* was older than *Alarm* and if either was to get through it might be easier to catch *Lightning* again. If I could not catch her it was John's turn to win rather than Stewart's and victory for so old a boat as *Lightning* would be good for the class and maintain the value of second-hand boats. I decided to let John go, and tacked to cover Stewart. The wind was altering all the time very slightly; now that John was on the opposite tack to us any small alteration of wind which headed us on our tack freed him up on his and doubled the effect of the alteration on the relative positions of the two boats. This could work the other way, too, and might have favoured us, but as it turned out the next time we came together with John, he crossed ahead of us and was in the lead at the weather mark. There was little change on the next two reaches except that John opened out from us and rounded the leeward mark almost exactly one minute ahead. By now the tide had fallen slack and it was no longer necessary to go close inshore. We were going to have the whole sea for manoeuvre on the last beat to the finishing line. As we came to the buoy with one more mile of windward work ahead of us, I said to my crew, 'It is a pity we couldn't win, but it's rather nice that old John in his four-year-old boat should be winning it, don't you agree?' This released a tirade from Charles, the gist of which was that he had never expected to hear such defeatist talk from me, that it was absurd to say that the race was already won and why didn't we get on with it?

'But,' said I, 'you know as well as I do that with competent operators in boats as nearly even as these are it is not possible to pick up a minute in a mile.'

'Well, you'll never know whether it is or not if you don't try!'

And so we set to and tried as never before; and at that critical moment there came a freshening of the breeze. After a couple of tacks we were sitting full out to keep the boat upright and driving her as hard as we could go. At the same time John and Tom Paxton (who were a good deal lighter than the combined weight of Charles Curry and me) were beginning to look back over their shoulders – beginning to realise that the freshening wind could beat them yet. Perhaps they were getting a little tired too, and earlier in the year John's appendix had been removed, which may have been a factor. Whatever the reasons, there came a time when the impossible seemed almost within our grasp. Up till now we had been carefully covering Stewart as well, and John of course had been covering us – tack for tack. Now we decided to wriggle and try to escape from John's control. We took a couple of short tacks, got him slightly out of step and unbelievably we were through. On the next port tack, we crossed his bows by about five yards, tacked to cover him, killed him with our wind shadow and crossed the finishing line sixteen seconds ahead of him after a race of three and a half hours. Twenty-two seconds after *Lightning* came *Alarm*, closely followed by James Beale and Mrs Richardson.

It was a moment of immense elation. During the last minutes of the race there had been no time to savour the golden instant when we took the lead again. Now the gun had gone; the unbelievable had happened; at the fourth attempt we had won the Big Race. The years of failure had perhaps built it up to a disproportionate significance. But even now I look back on those few minutes after the finishing gun as among the most triumphant and utterly satisfying moments of my life.

Apart from his skill as a crew and his knowledge of sail trimming, Charles Curry had undoubtedly given me a touch of the whip at exactly the right moment. His part in our victory had been much greater than is usual for a crew.

Although, sailing in his own dinghy, he competed in most of the Prince of Wales's Cup races during the next twenty years, and won the Olympic silver medal in the single-handed class at Helsinki in 1952, it was not until twenty-two years later when the Big Race was again at Lowestoft that he finally won it, by the even narrower margin of four seconds.

Later that summer came the regatta at Torquay, to commemorate the Coronation of King George VI. Stewart Morris had been invited to sail the six-metre yacht *Coima*, and the owner, Bill Horbury, suggested that he might select his own crew. Stewart asked me to join him as mainsheet hand; the rest of the crew consisted of the owner and Sandy Wardrop-Moore with Charles Curry as foredeck hand.

The regatta was chiefly memorable for some spirited races

between *Coima* and *Noma IV* sailed by His Royal Highness The Crown Prince Olav of Norway. In the long run he was too good for us and carried off the six-metre cup, but we were second by three-quarters of a point in seventy at the end of the fortnight's racing. We believed that a beautiful little red spinnaker carried in the Norwegian boat, and known as Monsieur Stalin, was chiefly responsible for our downfall.

A minor crisis arose during this regatta. We had devised a new gadget which we believed would improve the six-metre's performance on a close reach: it was a light spar which was fitted on to the forestay and into the clew of the jib, following the mitre seam and holding it out in a flatter curve when the sheet was eased. This device we had christened 'the Google' but before using it we thought it essential to discover whether our distinguished foreign competitor was prepared to allow its legality; although we believed it to be legal we must obviously be careful to avoid an international incident.

With this in view I remember going over in a motor boat one morning, before the racing started, to call upon His Royal Highness who was staying on Tom Sopwith's yacht *Philante* (which was long later to become his own Royal Yacht, *Norge*). I outlined the principle of our device and told the Crown Prince why we thought there was no rule to prevent it. There was a very slight pause and then, very firmly: 'I do not think it is legal and if you use it I shall immediately protest.'

The Google was dead. We were sad about it at the time, but in retrospect I am satisfied that the Crown Prince was right. It was an awkward gadget anyway.

From *The Eye of the Wind*

39

The Trapeze

As the summer of 1938 approached I began to think of our new sailing plans. John Winter and I had long had a theory which we had now decided to put to the test.

We believed that the crew rather than the helmsman should be responsible for the tactics of a dinghy race. The helmsman's attention should be exclusively occupied with watching the luff of the jib and sailing the boat, especially to windward. The crew was in a very much better position to look all round at the other competitors, assess the tactical situation and decide when to tack. Neither of us had ever had a crew we could entirely trust to do this, so we had sailed our boats by feel while we ourselves looked around to see how the tactical situation was developing.

There was the additional advantage in the new idea that, in sitting out, the helmsman uses rather different muscles from the crew; if we sailed one round each of a long and gruelling race, by changing over we should be much fresher, and therefore much less likely to make mistakes or to be clumsy in our exhaustion. Some years before we had planned to join forces – become joint owners of a boat in which we could put our theory to the test. Until this year there had been one difficulty about the plan. I was unwilling to join forces with John until I had won a Prince of Wales's Cup on my own account. My pride demanded first a success without John, the winner of 1934, to show me how. But having finally won the Big Race in *Thunder* at Lowestoft the way was now clear, and we had commissioned Uffa Fox to build us a new boat. In order to make it quite plain what the new boat was and how the system would work she was called *Thunder & Lightning.* John planned to retain *Lightning* and I retained *Thunder.* If the idea did not work we had our escape routes ready.

We had also had another idea – the device which is now universally known as the trapeze. Some years before I had crewed Beecher Moore in his Thames Rater at Surbiton to which he had fitted a 'Bell rope' attached to the mast at the 'hounds' and one member of the crew hung on to this and was thereby enabled to lean much farther out than without it. Uffa, Charles Curry, John and I had discussed taking the invention a stage further by the use of a harness to be worn by the crew which could be hooked on to a wire hanging from the 'hounds', that is to say from the point of attachment of the main shrouds to the mast. In this way the crew would lean or even sit in the harness with his feet on the gunwale and his body horizontally stretched outboard. If it really worked this device would give enormously greater driving power to the boat than had ever been possible with toe-straps in the middle of the boat and the crew leaning out just as far and as long as his belly muscles would sustain him. The Canadian dinghies had also used a method of belaying the jib sheet to a cleat on a sort of breast plate strapped to the crew. Our harness would combine the two.

As well as this trapeze, we felt that the very light wooden centreboard, which I had tried out in *Daybreak* but which I had not dared to use in heavy weather, might be operated in combination with the trapeze method of keeping the boat upright. We hoped to keep her up with the trapeze and at the same time have the advantage of the greatly reduced all-up weight off the wind. *Thunder & Lightning* was completed only very shortly before the Prince of Wales's Cup Week, which that year was at Falmouth. As a result there was no chance to try out the trapeze in realistic conditions. First, it had to be kept secret, or moderately so, and secondly, during the early races of the week the winds were quite light.

On the evening before the big race John and I sailed out from Falmouth Harbour, round the corner into Carrick Roads and there, safely out of sight of all our competitors, we tried out our new device in a very light wind. If I sat out to leeward and pulled the sheet in tight, it was just possible for John to go out to windward on the wire. The device seemed practical. How would it affect our performance? Of this we still had no clue.

There was in our minds no doubt that the trapeze was legal. Outriggers were not allowed, but nothing solid protruded from the boat in this system, and a wire hanging from the mast could not possibly be described as an outrigger. However, so that there should be no doubt of the legality if we used it, on the night before the race, when it was too late for anyone else to apply the system to their own masts, we leaked the information in the bar. There were cries of derision. We even showed them the harness and breast plate, which had been made for us. It brought loud and ironical laughter, and the cry, 'No big race has ever yet been won by a gadget.'

The day of the race again dawned utterly calm. We had to consider whether or not we should put in our light centreboard. The trapeze, its concomitant, could be used at will, but the selected centreboard had to be fitted before launching the boat. Was it to be the one made entirely of wood, or the one ballasted with fifty pounds of metal? We dashed by car up to the headland and looked out over Falmouth Bay. The race was to be held in the open sea and there was the merest cat's paw of a breeze lying across the face of the water. Sitting on a seat overlooking the bay was an old man in a navy blue jersey and peaked cap. 'What do you make of the weather?' we asked. 'How much wind will there be by midday?'

'Well,' said he, 'I think that'll blow. You wait till the sea breeze sets in and you'll find you've got all you want out there.' This was a most encouraging prophecy, but how good was the prophet? We were disposed to be convinced that none was better.

We hurtled down the hill, put in the wooden centreboard and were towed out of the harbour, through Carrick Roads and round the headland into Falmouth Bay eating our sandwich lunch as we went. By the time we reached the Committee boat a light breeze was blowing and it was evidently freshening all the time.

There was a big fleet – more than fifty boats – and the usual build-up of dry-mouthed tension. Just before the start it had become a planing breeze, We were carrying *Lightning*'s old mainsail, a famous sail which had been borrowed by Stewart Morris for one of his wins in *RIP* and used by John for his win in 1934. A sail which had won Prince of Wales's Cups already, and been second in two more, had certain magical quality quite apart from its proven excellence as a power unit. It was getting old now, and perhaps more suitable for heavy weather; through stretching it was flatter than it had been, but it seemed just right for a good whole-sail planing breeze.

I was to sail the first half of the race and I did not make a very good start. Immediately to windward of us and slightly ahead was Robert Hichens in the latest home-made dinghy *Venture II*. We were both on the starboard tack and in a matter of moments John had belayed the jib on a cleat and was out on the trapeze. Standing horizontally out from the boat with his feet on the gunwale, he was a startling sight even to me. To the other competitors the spectacle was irresistible. At an important time a great many of them gave their attention to our trapeze at the expense of sailing their own boats. Robert Hichens was now almost directly ahead; his crew looked at John with amazement, drew Robert's attention to him and or a critical ten seconds Robert sailed his boat 'off the wind', which allowed us to luff across his wake and get our wind clear. At one stroke we had escaped the consequences of my bad start and with John's weight keeping the boat much more vertical, with much less effort than any of the others, we forged ahead and rounded the weather mark by thirty seconds. But what about the jinx? Still in all the history of the Prince of Wales's Cup no one had ever won the race who had been first at the first mark. I said something about this to John, whose sharp reply was, 'There's got to be a first time.' Stewart was lying second for while until he was overtaken by Colin Ratsey, sailing brilliantly in a very broad-beamed ugly-looking dinghy called *Hawk*.

We finally finished nearly four minutes ahead of him, with Stewart third. Although it was satisfactory to win and as a race it was exciting in the extreme, it could not quite compare with the thrill of my first success in *Thunder*, the year before. The holder of a Championship has more to lose than to gain. If he wins, so he should; if he does not, eyebrows are raised.

But as soon as the race was over the trouble began. Was the trapeze legal or was it not? When it was pointed out that the rest of the fleet had had an opportunity to say whether or not they thought it illegal before the race and had not done so, nobody was ready to enter a protest against it after we had won. All that was left was to say that in future it must be banned and, on the principle of setting a thief to catch a thief, I was asked by the Yacht Racing Association's Dinghy Committee (of which as a Cup-winner I had now become a member) to draft the wording of a rule which would ban our exciting new invention.

I am still sorry about this decision. It may be that it would have radically changed the fourteen-footers, yet here was a system of keeping a small dinghy upright in strong winds which was eminently enjoyable, required no very great skill, but looked spectacular, appeared to have no danger, and reduced the compression strain on the mast. Most important of all it made the body weight of the crew a little less critical, because it enabled a light crew, for example a girl, on a trapeze to compete on even terms with a heavy man in wind strengths up to fifteen knots, whereas previously the light ones had been at a disadvantage in any wind above ten knots. All these advantages were there for the taking, but because of a prejudice against what appeared to be too difficult and acrobatic and what was imagined to be dangerous, and perhaps most of all because we had won the Prince of Wales's Cup by its use, it was outlawed and did not in fact return until the design of the Flying Dutchman seventeen years later. Now it is also carried in the 505 class and one or two others as well; and greatly enhances the enjoyment of sailing on a hard day. It is tremendously exhilarating to stand out, comfortably supported by the trapeze, almost horizontal and skimming low over the waves.

It is sad that a handful of people who did not have the vision to see this should have outlawed the trapeze for so long.

From *The Eye of the Wind*

PART IV

HAPPY THE MAN

40

The America's Cup

To me the words 'America's Cup' mean the most publicised and spectacular failure of my life. To Bob Bavier they mean the exact opposite.

Down the years those two words have meant quite different things to different people: to Sir Thomas Lipton a life-long dream which obstinately refused to come true; to the Earl of Dunraven a long tale of acrimony and international dispute; to Sir Tom Sopwith an agonising near-miss that could not be recouped; to Harold Vanderbilt a glittering record of successful defence.

The America's Cup is unique as the only top prize of a sport which has never changed hands in more than a century, and still retains front page news value. As one of the world's few still-unconquered Everests it continues to capture the imagination of millions, in spite of the ignominious defeats so often meted out by its defenders.

Constellation's supremacy over *Sovereign* was one of the most convincing in the Cup's history. The yachts' names in this context include not only the hulls, the spars, the sails, the instruments, but also the helmsmen and the crews, and there will long be argument on which of these factors were primarily responsible for the result.

A part of the answer will be found in *A View from the Cockpit*, and only one aspect of the story does not seem to me to emerge clearly enough, which may not be so surprising, because the author Robert N. Bavier, Jr is a modest man. The point should not be missed that *Constellation* was sailed with superlative skill by a great natural 12-metre helmsman – probably the best in the world.

Bob Bavier is that rare combination – a brilliant helmsman who can write. Bob manages to maintain the excitement and suspense all through the book, and even into the anti-climactic climax. The result cast a rosy glow over all that went before in *Constellation*'s hectic summer preparation.

It is sometimes difficult to prevent that same result from giving a bitter taste to the whole of *Sovereign*'s equally hectic summer. And yet until the last four races of the season, things had gone well enough for *Sovereign*. Off Portsmouth, England, and Newport, Rhode Island, *Sovereign* had battled with her near-sister ship *Kurrewa V* for the honour of challenging and had won by the narrowest of margins after a summer of wonderfully exciting racing. Whatever happened we should now at least be second in the America's Cup!

That the chances of winning were not good was first borne in upon us when *Sovereign* was too often beaten in early practice races by *Sceptre*, the badly defeated 12-metre of the previous challenge. Further writing appeared on the wall when we first saw the American 12-metres at Newport and were able to compare their hulls and their masts and their sails with our own. Their refinement at once appeared much greater than ours, but at that stage there was nothing we could do about it, and there was still a tiny hope that our assessments were wrong, that the indications of major hull improvement derived from the tank tests on the hull model were right, and that Olin Stephens's *Constellation* was not the startling advance on all previous 12-metres that her battles with *American Eagle* suggested.

'By 12.15 on 15 September 1964,' we used to say, 'we shall know the answer, for by that time we shall have been going to windward in close company with *Constellation* for a quarter of an hour.' And when that time came, our worst fears were confirmed. We could not sail as close to the wind as *Constellation* by several degrees – perhaps as much as 5°. The rougher the sea, the greater the discrepancy.

Having discovered that what we believed to be the best compromise between angle on the wind and speed through the water was simply not good enough, it seemed only right to try something different. We tried sailing fast and free, and we tried pinching up and accepting a slower speed. These experiments of desperation only put us further astern, but they had to be tried.

With the wind behind her *Sovereign* was no slower than *Constellation*, though bad judgement, especially in the selection of the right-sized spinnaker, lost us time on most of the runs – as much as seven minutes on one of them. For some sad reason we could not break loose from a belief that a huge parachute which had pulled us splendidly in calm water could be made to do the same in the steep and confused seas caused by the spectator fleet. But it was the effect of these seas on *Sovereign*'s windward performance which seemed to be the prime cause of our downfall. We seemed to hobby-horse much more that *Constellation*, and each time we pitched forward the boat seemed to stop in her tracks. How much this had to do with the shape of her hull, or with the mast, or the sails, or with the helmsman's inability to sidle her through the seas will never be known. One thing is certain: it had nothing whatsoever to do with her crew; I am convinced that apart from steering and decision-making the crew of *Sovereign* was every bit as skilful and competent as the crew Bob describes so vividly in his book. They were a splendid team all through the summer and never more so than in the last dark days of defeat.

So we were beaten – badly beaten – a combination which failed miserably to compete with a superior combination, but after all we had done our best. The series had been conducted without a single unhappy incident or protest. Much had been learnt about 12-metres; their design, their sails, their handling.

New friendships had been made and old ones confirmed. Even as we sailed back into Newport Harbour after the last race of the Cup series I knew that for me the effort had been worth while. Tony Boyden, *Sovereign*'s owner, has told me that he felt the same, even on that last bitter evening.

Foreword to *A View from the Cockpit*

41

The Spirit of Adventure

I believe that the spirit of adventure is not necessarily a craving for physical thrills strung together in rapid succession. It is an attitude of mind which can be applied to any field of human activity. It is the capacity to delight in the unexpected, to be undaunted by misfortune, to accept the challenge of 'come-what-may'.

You may find as much excitement catching butterflies in Surrey as hunting tigers in the Indian jungle.

The age of adventure, some say, is dead. But that is rubbish! The adventurous spirit and the opportunities to exploit it are no fewer than they were in the days of Sir Francis Drake.

Every day we read about some new trail being blazed, perhaps by a scientist, or a test pilot, or an explorer (and look at our twentieth-century Sir Francis – Sir Francis Chichester!); and before long the supreme adventure of space travel will be a practical possibility. But on our own earth there is as much scope for prospecting and development, as many openings for adventure, as ever before.

Some years ago I was in Canada, for instance. Canada is a wonderful land of opportunity for anyone with adventure in his blood. The country is opening up at a terrific pace. Throughout the north new ventures, new projects, new outposts are sprouting up with hundreds of interesting jobs going for people who don't mind living in the wilds with Eskimos and Indians for their nearest neighbours.

I believe the spirit of adventure burns inside most of us, whatever our job or walk of life. It merely varies in degree and expression.

I shall never forget a soldier I once met in the Normandy Bridgehead in 1944. I was in the Navy, and it was one of my jobs to investigate a harbour basin on the River Orne and examine some half-scuttled enemy ships. This meant a patrol into No-man's Land, crossing a large reed bed which was thickly mined, and being sniped at.

The Army detailed a young soldier to guide me. Clad in camouflage kit, he was in his element. He stalked through the reeds as stealthily and skilfully as a cat, enjoying every minute of it, yet, although he stalked like a born hunter, he was a Cockney lad born and bred in London.

How can a boy satisfy his appetite for adventure today? First I would advise him to get interested in something – it does not matter what it is. It may be something that other people will think trivial and boring. It is only vital that *he* should think it important. In my case I have selected things which happen in the country, in wild places, and especially things to do with natural history.

The one great thing about adventure is that you must have what you think is a worthwhile object – something to believe in, something to achieve. It may be a part of what you may be inclined to dismiss as 'only a hobby'. But that hobby may turn out to be your greatest contribution to the world.

So I would tell my adventurous friend to find some problem that needs elucidation. Then I would simply say, 'go off and discover the answers'. He will be adding something, however small it may be, to the sum of human knowledge.

If he has no money it will be more fun finding out. He will need to exercise a little ingenuity, and maybe that will add zest to the task.

Above everything else he must be tremendously excited about his mission. It must become for the time being (and perhaps for longer) the dominating aim of his life. My great key to the enjoyment of living is enthusiasm. Nothing worthwhile can be won without it. Those people who 'couldn't care less' would get a lot more fun out of life if only they could exchange their cynicism for a little enthusiasm.

Finance is the explorer's biggest headache. But money is not essential. The lack of ready cash would not prevent the true explorer from travelling to the other side of the world. The really determined person would be prepared to rough it. He would be willing to stint himself for two or three years in order to scrape together his fare. Or he may decide to work his passage, taking on the toughest of hard jobs (unfortunately, due to union restrictions, this is easier said than done these days). An adventurous friend of mine once worked his way across the Atlantic by looking after an elephant through the voyage on a cargo boat.

Some people appear to think that exploration is a sort of thrilling game in which the traveller tries to cram as much excitement as possible. But it isn't like that at all. The experienced traveller plans his trip well ahead and to the smallest detail. He reduces the possibility of excitement to the smallest minimum, knowing full well that he will get all the thrills he wants when things go wrong. Our trips have been planned with extreme care, for while the Arctic, for instance, is normally a friendly and peaceful place, it can change suddenly if anything goes wrong. Fortunately, nothing serious has gone wrong on any of them.

My worst experience, I think, was a rather nerve-racking flight down a winding river gorge less than a mile wide with thick cloud covering the tops of the cliffs on both sides and a heavy snowstorm, which suddenly reduced our visibility to about 200 yards. That was on our way into the Perry River country in Arctic Canada in 1949. We did not look for that adventure. It was accidentally thrust upon us.

It is obviously foolish to risk one's life unnecessarily. But an explorer, like adventurers in other fields, should I think be so enthusiastically obsessed by his mission that he will eagerly make sacrifices to reach his goal.

In 1936 I was exploring the Caspian Sea, hunting for the rare and beautiful red-breasted goose. One day during the search I thought I had found it. But a long stretch of soft mud lay between. In order to make certain (and at that time it was the most important thing in the world to me to find out) I should have to make a fairly hazardous passage across this mud. I remember thinking it out quite carefully and realising that the whole object of my expedition depended on this stretch of mud, and I remember wondering what sort of an idiot I was when I decided to have a go.

Actually, I got stuck in the middle and began to sink in. But fortunately it was not my first experience with soft mud. You have to know when to get out and cut your losses. You have to know when the situation is desperate, and when to throw yourself down and try to roll clear. I knew the danger so I threw myself down flat in the shallow water and succeeded in rolling clear – smothered to the eyebrows in slime. But I was clear on the right side. I could stalk closer to the birds, and did so, only to find that they were not red-breasted geese after all!

But exploration is only one of many fields of adventure. And it is not always necessary to leave home in search of excitement. Adventure is all around us, in art, perhaps, or science, or sport. You may find it in a gentle quiet form, a sort of arm-chair adventure, as I have done, for example, in painting and writing about birds. Or you may find it in rock-climbing, or gliding or sailing dinghies. There is no limit, because adventure is an attitude of mind which can colour all our activities.

42

Mr Spriggs and the Crane

'Do you know,' said Mr Spriggs, who was the district registrar's clerk, 'that it's the child next door's birthday?' 'Oh', said Mrs Spriggs, and there the matter dropped. She didn't even say 'Do you mean that horrid little brat that screams all the time?' nor did she say 'How do you know?' She just said 'Oh', because she was not interested in children. It was a source of unhappiness to Mr Spriggs, for he believed that he would make an ideal father. He had a way with dogs and other animals and he was sure that he had a way, too, with children, but – well, there it was, Mrs Spriggs was not interested in children.

Mr Spriggs glanced across the breakfast-table to his wife. She was a good wife, she had made him a nice home. His eyes wandered round the room. The pictures were nice – they were in nice frames and the wallpaper was nice: the lace curtains were nice and the little potted fern too; Mrs Spriggs had heard there was some joke about having an aspidistra, so she had a feathery fern instead. It was all very cosy, the perfect front room of the perfect suburban villa.

And to the outside world, to the other inhabitants of Mountview Avenue, Mr Spriggs seemed to be as happily married as any man could be.

Indeed so he was. But there was always just that one thing that weighed upon his mind – however, Mrs Spriggs was not interested in children, so that was that.

It was by pure chance that Mr Spriggs had happened to see the registration of the birth of the little boy next door. He had been turning the pages of the register the day before and his eye had caught an address which was so near to his own that he had read the particulars 'born 15 June 1928' – so the child would be three.

And now it was his birthday. As Mr Spriggs put on his drab raincoat, which he always wore to go to the office, no matter if it were midsummer, and his woolly grey soft hat, the seed of an idea was sown in his mind: as he passed down the narrow hall and opened the front door it began to germinate: and as he reached the garden gate at the end of the little gravel path, which he had so carefully rolled the night before, it had blossomed forth, for looking back he caught a glimpse, up the little yard which separated the house next door, of a pair of tiny bloomers hanging from the clothes line and waving in the wind.

Mr Spriggs left his office half an hour early that afternoon. He obtained special leave, for the seed was beginning to bear fruit in his mind. So he made his way in the direction of a toy shop which he dimly remembered to have seen from the bus.

In his pocket he had a ten-shilling note, but he couldn't possibly spend more than half of that. Five shillings would be the extreme limit, and even then if his wife should ever hear of it – Mr Spriggs shuddered, for Mrs Spriggs abhorred sentimentality.

A few minutes later he reached the shop. He paused before going in. The window was full – so full that no space could be found anywhere, behind the seven-foot-high plate glass, that did not hold a toy; there was everything there from a model motor that could be pedalled like a bicycle, to a beetle on a roller which walked slowly along if you ran it backwards first.

Mr Spriggs wondered how one began. His inclination would be to look round until he saw something nice – but he mustn't plunge in too soon without a moment's thought. His wife would of course classify the possible birthday presents into:

(a) Mechanical toys – cars, ships, aeroplanes;
(b) Sentimental toys – teddy bears, etc., of purely emotional value;
(c) Intellectual toys – Mr Spriggs couldn't think of any, but he was sure that would be one of them – it was one of his wife's 'words';
(d) Dangerous toys – etc., etc.

And then she would cross-classify with suitability to age, and finally eliminate with price. That is to say she would do all that *if* she ever bought a toy, which of course she wouldn't.

Mr Spriggs started to classify:

(a) Mechanical toys:

Model trains – No. 0 gauge – circular track in 10 pieces – 1 set points – engine and three carriages – engine looks like nothing on earth.

No. 1 gauge – track in 15 pieces – 2 sets points – engine still looks like nothing on earth.

No. 2 gauge. Oval track in 20 pieces with circular siding – how could he remember the particular merit of each one? It was impossible. He started to classify more freely and in less detail, but he felt he was skimping it, and went back to No. 3 gauge. Now of course his wife would write the details down – after all, he only had to copy them off the lid of either of the two show specimens. Where was his pencil? He tried each pocket, oh well, if he hadn't got a pencil – and anyway he didn't want to buy a train.

Mr Spriggs took the plunge. He opened the door and shut it again rapidly behind him, for a powerful electric bell nearly deafened him. It sounded just like his alarm clock!

There was no need for it either, because an imposing-looking woman was sitting at the cash desk, who could not have failed to *see* even the unobtrusive Mr Spriggs enter. She called 'Miss Harris – *SHOP*', and went on reading her novel.

A moment later Miss Harris appeared, Mr Spriggs was a little shy. It was two against one. However, he decided to broach the subject confidentially, so he said:

'I wonder if you can suggest a suitable birthday toy for a little boy of three.'

'Have you thought of a train?' said Miss Harris.

'Yes – I don't think I want a train.'

'Well, what about a tricycle?'

'Oh,' said Mr Spriggs. 'Well, I wasn't thinking of anything quite so expensive – I don't want to spend more than four or five shillings at the most.'

'Oh, I see! Well, how about soldiers?' suggested Miss Harris.

'Can I see some?'

A moment later she brought a long flat box. Inside were two rows of scarlet-coated guardsmen in heroic attitudes, all made of lead, and painted. They were tied round the neck and round the feet by a strand of white cotton to a sheet of cardboard.

Mr Spriggs didn't like them much; they looked so undignified when their heads came off, he remembered.

Now he would give Miss Harris one last chance . . .

'Have you thought of an intellectual toy?' she suggested.

Mr Spriggs gasped – and then asked what kind of a toy *was* intellectual.

'Oh, a rag book or some kind of a puzzle.'

'No, I don't want an intellectual toy, thank you,' he said.

Suddenly Mr Spriggs bethought him of a new method. So far he had tried 'classification' and 'confidence in the saleswoman' and both had failed him. Now his mind travelled swiftly back forty years to his very earliest memories. What sort of toys had he had? There had been a lovely singing top.

'Have you a singing top?' he asked.

'No, I'm afraid we haven't but we could get one in by next week,' said Miss Harris.

Again Mr Spriggs thought. Dimly he saw two children, the one lowering various articles from the first-floor window to the other in the garden, by means of a crane.

'Have you a crane?' he asked.

'Well, there's this one,' and she showed him a contraption of tin made in two halves which were clipped together.

'Have you a better one?'

'No, I'm afraid we haven't,' she said.

'Look in the second drawer on the left of the string box,' said the woman from the cash desk in a voice of thunder. Miss Harris looked, fumbled in the drawer, and then produced a large box. On the lid, in bright colours, was a lively scene from Southampton docks: a crane was unloading cargo from a huge steamer, dockyard hands were running hither and thither carrying bales and baskets on their heads, and goods trucks were waiting on a railway siding to receive their loads. Mr Spriggs drank in the scene and then Miss Harris opened the box. Within lay a beautiful crane. Mr Spriggs lifted it gently out. If you turned the handle it pulled up the string. And not only that, but it was geared down so that you turned fast and the string came up slowly. At the end of the string was a weight and a hook. Mr Spriggs set it up on the counter and lowered the hook to the floor.

'These are loads which can be pulled up,' said Miss Harris, pointing to three little sand-bags which were in a separate compartment in the box. Mr Spriggs put one on the hook and then hauled it up.

'How do you turn the whole crane?' he asked.

'With this handle. The whole platform revolves,' said Miss Harris.

Mr Spriggs tried the handle. It didn't turn.

'I think it's jammed,' he said.

Miss Harris tried it, but still it didn't turn.

'Have you another one?' asked Mr Spriggs.

'I think so' – again Miss Harris fumbled in the drawer and emerged with another 'scene at Southampton docks' which contained another crane.

Mr Spriggs tried its handles. They both worked.

'Good,' he said, 'I think I'll take that one.'

'Anything else?' asked Miss Harris.

'No.'

And Miss Harris began to pack up the crane in brown paper, and had just folded in the ends, when Mr Spriggs suddenly remembered the financial part of the transaction.

'How much is it?'

'Eight and six,' said Miss Harris.

'Oh,' said Mr Spriggs. Of course it was a beautiful crane, but – well, he had said he would only pay five shillings. And what *would* his wife say if she heard he had spent eight and six, and it would be worse if she heard that he had meant only to spend five shillings, but had been weakminded. No, he wouldn't have the crane. How did one break the news gently, when it was already packed up?

'I'm – I'm not sure,' he began, 'that it's exactly what I want after all.'

But Miss Harris had turned away to the string box and had not heard him. And at that moment Mr Spriggs's eye lit upon the clock. It was half-past five. The child would be going to bed in about half an hour. He must hurry.

'Pay at the desk, please,' said Miss Harris.

'One and sixpence change,' murmured the imposing-looking woman, and a moment later Mr Spriggs was in the street with a huge brown-paper parcel under his arm.

After five minutes' walk Mr Spriggs became one of the crowd that stood on the pavement waiting for the bus. And when the bus came he was one of the crowd that pushed and shoved to get in, and when the bus went, he was one of the crowd that remained on the pavement!

But he caught the next bus, and climbed up to the top. There was no room there and when the conductor came up, he sent Mr Spriggs down because it was strictly forbidden to stand on the top of a bus. Mr Spriggs went down and stood inside the bus. He was very hot and bothered, for you must not forget that it was midsummer, and all the time he clutched, under his arm, his huge brown-paper parcel.

At length he found a seat and wedged himself between a fat old woman and the end of the bus. First he laid the parcel across his knees, but the old woman complained that the corners were too sharp and were sticking into her; so then he put it longways, but at the next stopping-place another woman caught her parasol on it as she got out, and ripped some of the paper off. The passengers twisted their heads so as to look at the scene of industry in Southampton docks, at least, that part of it which was laid bare by the rent in the brown paper. Mr Spriggs turned it round and packed it down again as best he could.

Suddenly the conductor said, 'Addison Road.'

Mr Spriggs looked up quickly: 'Have we passed Mountview Avenue?' he asked.

'Last stop,' said the conductor, unconcerned. 'Any more fares, please!'

So Mr Spriggs got out and headed back towards home. Ten minutes later he turned up the avenue, the brown-paper parcel held firmly under his arm.

As his house came into view, he saw, to his horror, Mrs Spriggs leaning on the garden gate, talking to a friend. This was an unforeseen difficulty and Mr Spriggs decided he must wait until she had gone in, for it would have been impossible to reach the house next door unobserved. So here another ten minutes of valuable time was lost before he finally reached his goal and knocked upon the door.

Now was the great moment, thought Mr Spriggs; how surprised they would be. They would never guess how he had found out it was the child's birthday. They would ask him and he would say 'Aha', or might even say, 'A little bird told me.' Mr Spriggs loved that kind of surprise.

Presently the mother of the child opened the door. 'Mr Spriggs!' she said. 'Come in,' and she led the way to the front room.

On the floor sat the child playing with some bricks and around him were the grown-ups: the father of the child, an old lady and another girl.

'Auntie,' said the mother of the child to the old lady, 'this is Mr Spriggs from next door; and this,' turning to the other girl, 'is my cousin Flossie, Mr Spriggs.' Flossie smirked and said, 'Pleased to meet you!' Then there was rather an awkward silence, before Mr Spriggs produced the parcel.

'I've – I've brought just a little something, a toy that I thought the kid might like.'

'How nice!' said the mother.

Mr Spriggs felt things weren't going right so he began feverishly to unpack the parcel. The child meanwhile continued to pile brick upon brick in the construction of a monstrous tower.

At last the crane emerged triumphant from the welter of brown paper, string, and Southampton dock, and was placed proudly on the floor by the child's side. The child glanced at it, kicked it out of the way with his foot, and went on with the brick tower, whilst Flossie giggled and the aunt sat very straight in her chair and looked on.

Mr Spriggs collected the crane and produced one of the sand-bags; then, kneeling on the floor, he set up the crane on its box and began to demonstrate its possibilities – but the child turned back to the brick tower. Mr Spriggs produced another sand-bag and lifted it too from the ground to the level of the box. Here he unhooked it – it rolled – it fell to the floor and rolled further. Mr Spriggs turned quickly to pick it up, his foot caught the box, the box caught the bottom of the tower, and a moment later there was a resounding crash and the floor was strewn with loose bricks. Sitting amongst the ruins was the child crying bitterly.

Perhaps, thought Mr Spriggs, this was the moment for the crane, so he approached him again with it, but that only drew louder and more prolonged howls, and finally the mother came to his assistance and, picking up the child, carried him off to be comforted.

Mr Spriggs felt dazed. He said he thought he had better be going, and still under the aunt's gaze of now undisguised hostility he made for the door.

As he walked down the garden path his thoughts were very bitter. With children he had been a failure, his whole evening had been a failure, his eight-and-sixpence-worth had been a failure. His surprise had been a failure, for they had not even asked him how he knew it was the child's birthday. *He* was a failure.

He turned back up his own pathway and went into his house. He would have needed very little provocation to burst into tears.

Vaguely he remembered he had promised his wife to weed the garden in the cool of the evening. He went into the back yard and found a little trowel and then came out to the front again and started work on the herbaceous borders.

An hour later Mr Spriggs had reached the corner where, beside a laurel bush, two deck-chairs stood. The evening was warm and his back was aching so he sat down. Above the trees of the avenue swifts were chasing each other and every now and again screaming shrilly, a noise which jarred on his ear, almost as much as the memory of the afternoon jarred on his mind. A motor-bicycle roared past up the street and Mr Spriggs made a little grimace of annoyance. Then he heard the front door of the next house open and a man's voice say something – and a moment later a woman's voice answer. Ordinarily Mr Spriggs would never have eavesdropped, but tonight he felt at war with the world. He listened intently.

'Well, have you put him to bed?' asked the man, who was evidently already sitting in the garden.

'Yes,' said the woman, who had just arrived, 'at last I have; he would go on playing with that lovely crane.'

'D'you mean the one old Spriggs brought?'

'Yes – he's only gone to bed now without a fuss on condition that he can have it on the chair beside his cot. He's winding the sand-bags on to it.'

'Jolly nice of old Spriggs to bring it, wasn't it?'

'Yes,' said the woman's voice. 'I wonder how on earth he knew it was Georgie's birthday.'

'I don't know. I must ask him in the morning. Well, have you got the supper ready?'

'Yes, dear, come along in.'

Mr Spriggs felt that a peace treaty had been signed between himself and the world. So he got up and went indoors, for he wanted to tell somebody all about his afternoon. He found his wife in the kitchen and was going to tell her; but instead he said 'Oh, hullo dear, can I help you with the supper?' for he remembered that Mrs Spriggs was not interested in children.

From *Cornhill Magazine*

43

Britain's Canals

It has been a primary task of the Inland Waterways Association, of which I was one of the first Vice-Presidents, to emphasise that today the waterways have functions of ever increasing importance which were virtually undreamed of when most of them were built.

Prominent among these new functions is the service of pleasure and recreation. Here we have taken only our very first step into a new world. Already millions of our people receive holidays with pay by statute. One wonders where they all go, and if the accepted resorts are not becoming progressively more crowded. The answer is that at present only about fifty per cent of those receiving holidays with pay are actually leaving home. It takes more than a year or two to alter the habits of our conservative people. But in the end they will almost all go; and not all of them will want to go to holiday camps. Never does a day pass without the Inland Waterways Association's receiving twenty or thirty enquiries about how you can get 'a holiday on a barge', as most of them term it.

As the standard of living rises, a buying sequence comes into operation with remarkable regularity: first a house and furniture are purchased; then a car, TV and a washing machine; and then a man's wife may let him buy a boat. Another important factor is the increasing difficulty of finding pleasure in motoring: do you go pleasure motoring any more? There is in pleasure boating not only a wonderful form of recreation, but also an important new source of revenue for the navigation authorities.

And even here we are considering only the early stages. Before us seem to lie the alternatives of annihilation or a leisure state based upon automation and atomic energy. The former cannot really be planned for, and anyway I do not believe in it. The latter involves taking leisure seriously indeed.

Until I spent a month cruising the canals in a narrow boat I had no idea how utterly satisfying it could be; the gentle pleasure of travelling with your home, of seeing remote parts of the countryside from a new angle. Unlike the roads, the canals have for the most part escaped ribbon development. When they do pass through towns and villages, it is through the quieter parts, through the back garden, so to speak; parts you do not normally see if you pass along the façades of the main streets. I was surprised at the number of long, lonely stretches of canal in the highly populated Midlands. I was surprised by the fascination of Birmingham seen from the waterways. In a boat, too, there is that wonderful feeling of isolation and detachment – a spell which can scarcely be broken even by 'going ashore for a meal' in some restaurant or hotel. And then, of course, there is no telephone, surely one of the greatest advantages of all. I found it ideal as a holiday – peaceful, with an ever changing scene. But there is a more active aspect than this; there is the knowledge and skill required to navigate the boat and to work the locks. There are sudden groundings, narrow bridges which are narrowly missed, claustrophobic tunnels in which the boat may stick, engine failures and unpremeditated immersions. It was all a new world to me and one which is hard to imagine until you have seen it, indeed lived in it. It is beautiful and peaceful, and at the same time it fires the imagination and nourishes the Spirit of Adventure.

That much of that world survives at all is indisputably the achievement, almost unaided at the outset, of the Inland Waterways Association.

All who care about the waterways (and all who navigate them soon *will* care about them) will be gravely remiss if they fail to join the Association. There have been some major victories, but only battles have been won, and by no means the war. Support is still urgently needed. But there is no longer any question about public opinion's being in favour of retaining and restoring this great heritage. Otherwise the Association could not have gone so far in the time. It is said to take twenty years to alter the climate of opinion on any subject. It seems possible that this time it will be done in less.

From *Know Your Waterways*

44

Reflections

I have always been in favour of balloons – full sized proper balloons, preferably the kind in which you make an ascent. I have never been up in a balloon. But even captive ones have always been satisfying to me, perhaps because one of my earlier memories concerns a captive balloon. I can remember being driven into London, with my mother holding on to the seat of my pants while I leaned over the back of the motor-car to gaze at one of the great sausages riding above the treetops.

My life-long enjoyment of natural history had its origins in these early days, I think, because I can't remember a time when I wasn't interested in it. I was sent away from London's zeppelin raids to a little coastguard cottage at Sandwich, and there caught lizards and newts and frogs and watched birds, mostly waders; but one fateful day I saw a flock of wild geese and they have held an especial fascination for me from those days to these. I drew them all, too.

It had been my father's wish that I should be brought up with an interest in natural history: and as luck would have it I was quite keen on the idea myself. My schools were helpful about it too. Looking back, it seems to me that I spent most of my school days hunting for birds, catching rabbits, finding caterpillars and fishing – setting out with a fishing rod stuck down my trouser leg and the limp which was supposed to have got me off playing games.

—

'And what,' people sometimes ask, 'is the real attraction of bird watching? Why does it collect so many followers?' Well, for me its charm is that it's difficult, it's out of doors, it requires patience and an enquiring mind, and the objects of the study are not only interesting but beautiful. Having found something which commends itself to me for so many reasons, I'm afraid I follow it with what most of my friends regard as a disproportionate singleness of purpose, and those less charitable call a one-track mind.

I believe strongly that the pursuit of truth is a worthy aim: that it is worth finding things out for the sake of it. In short, I think that pure science is more important than applied science. It seems to me that human progress is only possible if the first consideration of the scientist is the advancement of knowledge. The application of that knowledge to the material requirements of man is by comparison a pedestrian affair, and much worse than that when more than half of it is devoted to destruction.

—

At Cambridge I had spent half my scholastic time being trained as a scientist and the other half as an artist. And still today I am half one and half the other, and no doubt for that reason, not very good at either. But the life suits me well enough, and living in the country, in the atmosphere of the marshes and the birds, is surprisingly exciting and adventurous – if that is how one looks at life. As a result of lecturing and broadcasting a good many letters come my way from boys and girls who seem to regard me as an authority on How to Live an Adventurous Life. I can only say that my life has seemed adventurous *to me* because that's the way I like it. I try to exploit the contrasts in life, I look for things to be exciting and fun and interesting, and then I find that they are. It's an attitude of mind, and you don't need to hit the headlines or go to Timbuctoo in order to have an adventurous life. For instance, I have found plenty on the canals and rivers of the Midlands in a converted narrow boat. The object was a waterborne lecture tour, combined with an opportunity to write without telephone interruption or the guilty stultification of long unanswered correspondence. The result was a month of exciting travel in which we covered 450 miles – from Gloucestershire up as far as Southport and Liverpool and back again. We went through 273 locks (working them ourselves), crossed the Mersey estuary, a 15-mile passage among the washes of steamers and tugs (the first converted narrow boat to make the crossing) and got stuck in the famous Harecastle Tunnel in the Potteries. The tunnel is nearly two miles long and our boat became wedged about the middle rather like a cork in a bottle-neck; we couldn't decide whether to push it in or pull it out. After more than six hours of struggle, loading bricks from the towpath, pushing, pulling, and jerking, we finally got her through, and emerged into a snowstorm.

I believe there are great possibilities for our inland waterways in the field of holidays afloat – possibilities which have so far scarcely been contemplated; for there are 2,000 miles of navigable canals and rivers leading from the centres of population right into the heart of some of our loveliest country.

—

Many people ask me what are the possibilities for themselves or their children in the field of natural history – what openings and opportunities there are for a career. The answer is that there are sadly few. A good many of these enquiries come from people who cannot fit into other fields in life, and in many cases would fit no better as naturalists, for it must not be supposed that there is any future in natural history for those who come to it as a last resort when all else has failed.

But all the same I could wish that the prospects were better than they are for those with ability and a special flair for natural history – those who are ready to work hard for a degree in the subject and want to follow it above all others as their career. The prospects are better than they were, with the greatly increased interest in wildlife, but they could be better still. I would like to improve those prospects; to create more opportunities for work in this field; I would like more people to be able to get as much pleasure from the study of nature as I have got, and to enjoy their lives as much as I am enjoying mine.

45

The Boy David

There had been a break in my early tea-time visits to Adelphi Terrace, to my godfather, Sir James Barrie, until a letter in April 1930:

> . . . having heard a glorious rumour that you are growing rather like your father I want very much to see you again. Could you come in and lunch with me on Wednesday or Thursday at one-thirty. I should be very glad if you could.
>
> Yours sincerely, J. M. Barrie

I went often after that and, after my training as a painter in Munich and London, I asked Barrie if I might paint his portrait. I knew that he had not previously allowed this to be done, and he told me that he did not think he should break his rule, 'but I would very much like you to paint a picture of this room, and, who knows, when you have finished it you might find me sitting in a corner of it'.

In January 1934 I found myself taking a canvas thirty inches by twenty, a box of oil paints and a small portable easel into the famous room in the top flat of Adelphi Terrace House. The centre of my picture was to be the open hearth piled high with wood ash, surrounded with various characterful fire irons, a pair of bellows with long handles, and a brass kettle. On the left was to be the leather-covered pouffe, on the right the ingle-nook containing the high-backed settle on which I hoped Barrie might sit as the picture advanced; and so in due course he did. I learned more of Barrie during the next few weeks than I had discovered in all the years of casual visits. He was no longer a godfatherly institution which rated a visit at about the same intervals as the zoo and the Natural History Museum; instead of a literary legend he was at last a real person. If he was older than he had been, so too was I, and our talk was easier than it had ever been. Often there were long silences which embarrassed neither of us. I painted and he sat gazing into the fire, enveloped in clouds of smoke from his pipe, and sometimes coughing alarmingly, so that I would suggest he smoked too much. Once I asked him whether he had ever planned a conclusion to his one-act play *Shall We Join the Ladies?* I told him that I had tried to work out plausible second and third acts for it, but had not been successful. He said, 'I never could make out which of them did it, and I don't suppose the mystery will ever be solved'; then, 'Most of them must be dead now anyway.'

I asked him whether he would write any more plays, but he said he was seventy-four and he doubted if he would. I said something empty like 'Oh, what a pity,' and there it might quite easily have ended, but evidently he wanted to go on thinking about writing. Suddenly he began to talk – excited talk of the days when he first came to London, of the Davis boys whom he had brought up after their father died, of how *Peter Pan* grew out of them; he talked about village cricket and about the Outer Hebrides and the origins of *Mary Rose*; he talked about the theatre in a friendly, loving way. I asked him how long it was since he had been to a play and he said three years. I said I thought this should be remedied; I had just seen a play which I thought he should see.

Soon after I began painting Barrie I was taken to Margaret Kennedy's play *Escape Me Never*, in which the Viennese actress Elisabeth Bergner was appearing at the Apollo Theatre. Ever since reading *The Constant Nymph* I had greatly admired

Margaret Kennedy and it seemed that she had produced, if not a perfect play, at least a perfect vehicle for the art of Elisabeth Bergner. Her performance as the young waif Gemma Jones moved me so much that I went to see the play again. It chanced too that my mother was instrumental in finding a house in Hampstead for Bergner and her husband, Paul Czinner. I went to the play a third time and a fourth.

When I first proposed to take him to the Apollo Theatre, Barrie was full of excuses. He said that he did not go to the theatre these days and did not think that he would enjoy it very much anyway. But I was not prepared to take no for an answer. I made a careful plan with Frank Thurston, his manservant. 'I'm coming to dinner with you on Wednesday, and if you feel like it, we'll go to the theatre afterwards,' I said, and left it at that. I had already bought the tickets and when the day came we dined together in his flat, delightfully as ever. After it Frank appeared with his coat. 'Time to go now, Sir James,' he said, and the pressure from both of us was too great. A few minutes later we were off to the theatre in my car. At the interval Barrie said nothing about the play or the performance. It would have been quite wrong to ask. We sat in silence. I scribbled a message on a programme and sent it round to Bergner saying that I had Barrie with me and might I perhaps bring him round to see her afterwards. Curiously enough there was no reply at the next interval; I must have given inadequate details of where we were sitting. But in spite of this I felt so sure that she would like to see him that we went.

In the last scene of *Escape Me Never* Gemma Jones breaks down into hysterical laughter from which I had always noticed that she was in tears at the final curtain call. When we found Bergner in the passage just outside her dressing-room she was still in tears. Without saying anything at all she took Barrie's hand and held it for rather a long time while fresh tears coursed down her cheeks. We hardly stayed with her at all. Barrie said, 'Will you come and have tea with me on Friday?' She said, 'Yes, where do I come?' I stepped in quickly and said, 'Never mind about that, I will fetch you.' And their first meeting was over.

I took Barrie back still in complete silence; it was not for me to break in on his thoughts. I was still not quite sure whether it had been a success.

Next day I was round at his flat and at work on the painting again. Barrie was pacing up and down as he so often did, smoking his pipe, coughing a little, his voice occasionally crackling, one eyebrow raised:

'You've unsettled me . . . '

'Oh dear,' I began to apologise.

'I told you that I had given up writing. Since last night I've decided to take it up again.' Then after a pause: 'I haven't felt like this about an actress since the first time I saw Pauline Chase.'

'What will the play be about?' I asked foolishly, but he did not answer. Thank goodness he hardly seemed to have heard the stupid question. In pigeon-holes in his mind he must have had many ideas, but the theme of his last play, *The Boy David*, seems to have come from Bergner. For the rest of the afternoon, as I painted, we reminded each other excitedly of the play we had seen. 'Do you remember where she says . . .?' 'Did you notice her hands when he came in in Act Two?'

I did not know until long afterwards when I read it in Cynthia Asquith's *Portrait of Barrie* that he had taken her to *Escape Me Never* on the very next night after I had taken him.

The picture was not completed for a week or two after that, but it was clear that the new play was not only secret but also taboo as a topic of conversation and I never returned to it. Nor was I ever allowed to be present when Bergner had tea with Barrie, although for some weeks I maintained the illusion that his flat was too difficult to find without an experienced guide. Twice a week, and sometimes more often, I would go to collect her in my car at Frognal and deposit her at Adelphi Terrace, returning for her after tea and driving her home again.

It was not for another two years that the new play finally

went into rehearsal and thereafter every conceivable kind of bad luck attended the production – illness, unfounded rumours, a string of postponements, and the Abdication, which burst upon London at the same time as *The Boy David*. When the play began its run I was abroad and by the time I returned it had been taken off. Four months later my godfather died.

From *The Eye of the Wind*

46

World Champion Skater?

I had never been to Berlin before, but I found my way by tram to the Sportpalast where the Skating Championships were being held. For three days I watched the skating. Winner of the men's title was Karl Schäffer from Vienna with a marvellously unhurried free-skating programme, with great slow three-jumps and single loop-jumps (Rittbergers, they were sometimes called in those days) in which he seemed to linger in the air. Women's Champion was the young Norwegian girl Sonja Henie, with a programme full of fireworks.

Watching the competition was the British teacher and stylist Bernard Adams, a gentle little man with white hair and moustache and black eyebrows. He had never himself been a great performer, but he was acknowledged as one of the greatest of all teachers of the art. In London I had known him well and had some lessons from him. Here in Berlin he made me an interesting offer. 'Loops or no loops,' he said, 'if you will give me your time uninterrupted for the next two years I will make you World Champion.' It was an interesting thought. Did I really want to give up everything else to become a champion figure skater? Sitting beside the rink in Berlin as the World Championships took place before us, this seemed to be an uncommonly attractive proposition. But when I had returned to Munich and was back at my painting, I thought to myself 'Perhaps not.' At the back of my mind was the thought that skating was an art, not a sport. Who ever heard of a World Champion Ballet Dancer? The objectives seemed somehow to be muddled. Sport or art, amateur or professional, all these seemed to confuse the issue, seemed to be far removed from the delights of the skating itself. Perhaps, too, Bernard Adams would never be able to get me over those loops, and even if he could, did I really want to give up two years of my life to this particular and rather limited objective? On balance I felt that I did not.

From *The Eye of the Wind*

47

Venice, 1939

Early in the summer of 1939 my mother and I were invited to stay in Venice with the pianist Prince George Chavchavadze. He had recently been married and was living in a palazzo overlooking the Grand Canal, with a private gondola and his own gondolier. This to me was the ultimate luxury and my mind went back to 1919 when I had learned to propel a gondola. Once more I tried my hand at it mounted on the *poppa* and found that after rowing a duck-punt the gentle art held no insuperable problems. Perhaps, if all else failed, I could find work as a gondolier.

George Chavchavadze was a beautiful pianist, then not yet perhaps at his highest peak but already wonderful to listen to, with a large repertoire. I made two drawings of him, one of which was rather successful in conveying the musician's far-distant look.

Besides being a very promising pianist George was charming and amusing. One of his parlour tricks was to convey a thunderstorm by facial expression, starting from the dim flickerings of distant lightning and finishing with the cloudburst overhead. He had set to music in operatic vein the legend on the current packet of Bromo toilet paper.

Our days were spent in much the same pattern, bathing before lunch at the Lido, going to see pictures or churches or the Doge's Palace or Verrocchio's magnificent equestrian statue of Colleoni in the afternoon and bathing again when we got too hot. And what pictures and architecture there were to see! From twenty years before I remembered the gondolas, and San Marco, and the glass factories, but I hardly remembered the pictures at all. Now I found myself bowled over by the vast frescoes of Tintoretto in the Scuola di S. Rocco and the church next door, and the huge 'Paradise' (supposed to be the largest picture ever painted on canvas) in the Doge's Palace. But oh, the Giorgione in the Giovanelli Palace, the Titians, the Veroneses and even the brilliant baroqueries of Tiepolo!

In the evenings George played for hour after hour as we sat on the balcony and watched the moonlight shimmer on the Grand Canal. So much beauty. But the 'handful of wicked men' were at work; European politics hung like a great cumulonimbus cloud over our summer.

From *The Eye of the Wind*

48

Journey to the South Pole

Ever since I can remember the Antarctic has been part of my background, but it was not until early in 1966 that I set off for the Antarctic for the first time myself. It was not just a sentimental journey, for I wanted to find out how different the forms of polar travel are these days. There are hundreds of people living down in the Antarctic now, and I wanted to know what they are doing, and what sort of life they live.

Right in the heart of London, just down the street from Piccadilly Circus, stands the bronze statue of my father carved by my mother. Later she copied it in marble for the people of Christchurch, New Zealand, 12,000 miles away, where it stands in quiet gardens by the banks of the River Avon, in that sleepy, and very English town that has become the traditional New Zealand gateway to the Antarctic. For me, too, it was to be a last glimpse of temperate and civilised surroundings before my long flight south to the huge continent which is larger than Europe and the United States put together.

I have always thought of it as all people must – as a place of terrible struggle and hardship, where each new generation of explorers has met the same bitter conditions as the last – and its own calamities.

Cape Evans on Ross Island, Antarctica, is nowadays just about ten hours' flying-time from Christchurch, New Zealand. And yet it took the men of my father's day some forty days to reach this same spot. Forty days of battling through heavy seas, cutting their way through the pack ice before their ship, the *Terra Nova*, could find an anchorage in the little bay at the foot of Mount Erebus. Only then could they start getting all their equipment ashore; the ponies, the dogs – all that they would need to make a base camp for their journey to the South Pole.

The resident skuas and their young must have resented the intrusion. My father describes in his diary how they dived on him, beating at his head with their wings. They still do. And at the side of their hut, at Hut Point, I found a carcass of a dog, preserved by the cold, lying where it must have died more than half a century ago. When I went into the hut I felt I was stepping back through the years. A New Zealand party has cleared it of snow, and tidied it up so as to make it look just as it did in my father's time; and they've done it so well, that for me this was a very personal moment, a sort of tangible page of family history.

Neat piles of tinned food fill the shelves, their labels stained with age and their contents, for the most part, just as good as ever. The ramshackle bunks were called the tenements – they are empty now, but once were full of the noise and laughter of some of the great names in British polar exploration: Birdie Bowers, Cherry Garrard, Mears, Titus Oates.

The hut is very cold now, and of course it's a sort of a museum – a bit like a stuffed bird compared with a living one. But you can feel that it was once a place of comfort and warmth and good cheer. The main mess room table, for all meals and general activities, ran right down the middle of the room. Beyond it, on the left, was the area that my father called his 'den'. There he sat each evening writing up his journal through the long, dark winter, with something of the loneliness and isolation of any ship's captain. Herbert Ponting's picture of my father sitting there is something I've grown up with. A big print of it hangs on the wall at my home in Gloucestershire.

It really was a most extraordinary feeling to be sitting there myself.

The leader of the scientists, and the closest friend my father had in the world, was Edward Wilson, always known as Bill. His bunk was in the corner, right across from the 'den'. My father wrote in his journal: 'Words must always fail me when I talk of Bill Wilson. I believe he's the finest character I ever met.' Within six months of that, they were both to die in the same tent – of cold, hunger, scurvy, and exhaustion.

The next part of my own Antarctic journey was in Wilson's tracks, along the south shore of Ross Island to the opposite end, to Cape Crozier – in search of penguins. Our journey took forty minutes each way. Wilson and his companions, Bowers and Cherry Garrard, man-handled their sledges through the perpetual darkness of mid-winter for five desperate weeks. Their sleeping bags froze stiff and flat like boards. At one point the tent blew away completely, and their clothes were so full of ice they were double their normal weight. Their destination was Cape Crozier, home of more than three hundred thousand penguins. Living out there when I went, carrying on the work which Edward Wilson began, I found an old friend, Bill Sladen, one of the world's greatest experts on penguins, who really loves Cape Crozier. From the social structure of this vast rookery you do get some ideas about territorialism and aggression, and all the things which can be applied, of course, to other animals, including human beings.

One of the astonishments of modern Antarctica is that man has already imitated the penguins and built a city of his own. It is McMurdo Base, built by the United States Navy and often called McMurdo City. This is the Antarctic headquarters of Operation Deepfreeze – with its own street lighting, a regular bus service, and cocktails mixed with the oldest ice in the world. McMurdo is really a frontier town. The first explorers have come and gone, and completely civilised living hasn't yet arrived. There's been scarcely any attempt yet to make McMurdo City beautiful, or to come to terms with the landscape. It's ugly, but it works.

On the hill above McMurdo the Navy has built the first nuclear power station in Antarctica – it's called Nooky Pooh. A two-year fuel supply for this station is no larger than a single oil drum, replacing the need for millions of gallons of diesel fuel every year. And with nuclear power, a chance to run a desalination plant. Believe it or not, in the midst of most of the world's ice, there's a water shortage! It's cheaper to take the salt out of the sea, and electrically heat the layers of lagging around the water pipes, than it would be to melt ten gallons of snow to make one gallon of water.

Just across the little harbour from McMurdo City is Hut Point – for me the biggest contrast in the whole of Antarctica. On one side, all the modern technological developments of McMurdo City, and on the other, the hut my father built in 1902. The walls are blackened with age and blubber smoke, and there's some mutton there, still hanging in the world's largest and cheapest cold store. The place gave me the feeling that the early explorers had a sense of environment and setting that seems missing today. Now, man has such powerful tools at his command he can't always be bothered with the niceties of meeting nature halfway.

Modern icebreakers work the channel through McMurdo Sound so frequently, keeping a channel free for the supply ships to get in, that some people think they may be changing the climate of Ross Island. It seems that the summer months are significantly warmer now than they used to be. Watching the US icebreaker *Glacier* at work, riding up over the sea ice and crushing it downwards with her own weight, made me wonder what the early explorers might have done with power like that to back them up.

The Antarctic isn't conquered yet, but as I watched the segments of shattered ice snake away from the bows of the *Glacier* in shimmering patterns of light, I felt that strength is now being met with equal strength. Ernest Shackleton wouldn't have had to watch his ship the *Endurance* held fast in the pack ice for 281 days, and finally crushed like a nut between the pressure ridges. He wouldn't have had to spend six months marooned with his men on an ice-floe as they drifted towards the open sea.

But with all this strength, the Antarctic is still a dangerous place. Almost every year of Operation Deepfreeze helicopters or aircraft have been lost. Hazards of weather and aerial navigation are probably as bad here as anywhere in the world. No crew leaves, even on the shortest journey, without an emergency survival kit. The risk of a ship failing to make port is pretty small now, but the risks for a plane down here are still considerable. And now a parachute rescue team has been started to bring immediate help to anyone stranded on the ground. For my father and his companions on that fatal march back from the Pole, rescue comes floating out of the sky – just fifty years too late.

While I was there I saw a sight that in a year or two I may count myself lucky to have seen. A team of huskies was being harnessed up for a sledge journey across the Ice Shelf. The dogs live outside in the snow the whole year round, often in temperatures of minus 70°F. And they eat nothing but a lump of raw seal every other day. It must be about the toughest existence that any animal has to endure on earth. For the first mile they go off at a tremendous pace, but it doesn't last. They soon settle down to a steady trot. If they feel like a drink they scoop up a muzzle full of snow on the move.

I travelled out to the camp on a modern motor toboggan. For me the going was crisp and easy. Along with ten Siberian ponies, my father's party, too, had a couple of motor sledges. He wrote in his diary: 'I find myself immensely eager that these tractors should succeed.' Well, the tractors broke down very early on the journey to the Pole, but he was absolutely right about their future. And absolutely wrong, some historians think, about his choice of ponies to pull the heavy loads to the bottom of the Beardmore Glacier. And wrong, too, not to use dogs for the later stages of the journey, up the Glacier and across the Plateau to the Pole, as Amundsen and his Norwegian party did. I don't think the decision was entirely sentimental. After all, the British expedition had to face the unpleasant task of shooting the gallant little ponies as they sank, exhausted, up to their knees in the soft snow. But they didn't think there was enough advantage in dog-hauling to justify having to drive their dogs to death.

The Norwegians started out for the Pole with fifty-two dogs, and ended up with eleven. Amundsen knew the precise day when each dog's usefulness ended for pulling supplies, and its usefulness began as a source of food. One of the Norwegians wrote afterwards: 'What shall we say of Scott and his party who were their own dogs? I do not believe men ever have shown such endurance at any time.' How different it all might have been if the motor sledges had been really reliable like the little toboggans I travelled on recently.

I arrived to the usual warm Antarctic welcome, and a mug of English tea, while we waited for the dog teams to catch us up. It seems a pity, but I don't think it will be long now before the husky becomes another nostalgic memory of the past.

As we settled down for the evening meal, I was very much aware how close we were that night to the place where Wilson, Bowers, and my father had made their final camp. Somewhere in the ice of that Ice Shelf, their bodies lie encased about 50 feet below the surface, and are gradually moving towards the sea.

Scott Base is the permanent headquarters of the New Zealanders near the southern tip of Ross Island, only a short distance from the American base at McMurdo. It was the nearest thing I found to what must have been the atmosphere and feeling of those early expeditions. These New Zealanders can turn their hands to anything; they are the Antarctic all-rounders, who can mend sledges, drive dog teams, take scientific measurements. Winter and summer, sections of the great ice pressure ridges have to be carted back and melted down into bathwater: no nuclear power station here. At the Base Post Office, with a post mark much prized by stamp collectors, I enjoyed the simple pleasure of posting letters to my family back in England – one to my wife, and one to my son, Falcon, so linking up three generations of Scotts from the Antarctic. Underneath my father's portrait we sat down to dinner. The scene might have been any dinner party in a hut in the Antarctic – like the last birthday party that my father was given at Cape Evans.

Ponting used to amuse my father's expedition in the evenings by showing them some of his magic lantern slides from Japan. Their place is taken today by modern pin-ups. By their very absence women seem to pervade Antarctica. Personally, I don't think it will be long before the girls will be allowed to come down off the walls and put on their fur-lined jackets like the rest of us.

The 800-mile journey to the South Pole in an American Hercules transport plane takes just under three hours from McMurdo. No one ever again will make the whole journey on foot, man-hauling sledges over the Ice Shelf and up the longest glacier in the world – the Beardmore. There, amongst the mountains and the glaciers, geologists have found coal, and other fossilised remains of a rich, tropical vegetation. No one quite knows why, but it seems that either the climate has changed dramatically at some time, possibly through a shift in the earth's axis, or else the whole vast continent of Antarctica has moved, has drifted south from the Indian Ocean.

Suddenly, we arrived at the South Pole, a tiny group of huts and radio masts, a cluster of dark specks on the vast, blank, white sheet of snow at the bottom of the world. Outside the aircraft the thermometer registered 65 degrees of frost. The snow under my feet, compressed into ice, was very nearly two miles thick; but it was the breathlessness that I noticed first, rather than the cold, because this was 9,000 feet above sea-level. Most of the Pole Station, or the Amundsen/Scott Station, to give it its full title, has been built beneath the surface, to protect it from the blizzards that sweep across the polar plateau with nothing to break their force. But, ironically, the greatest danger which faces the men who man Pole Station is not cold, but fire. The wind can turn a spark into a blaze in a few seconds, and if essential stores get burnt up in so remote a place you're in real trouble.

Below ground, with no sun shining, it's even colder than above. In the tunnel between the huts there is a constant 75 degrees of frost, though inside the huts it's comfortably warm – indeed, sometimes almost too hot. A group of about a dozen scientists sit out the winter darkness there, physically cut off from the rest of the world; they photograph the ionosphere, record cosmic rays from outer space; study the earth's magnetic field: and down this underground tunnel, a thousand feet long, below the ice, one of them had instruments that record an average of fifteen earthquakes a day. Some of the work has practical application, like the meteorology; a tangible return on the $27 million that the United States spends every year in Antarctica. But some of it, like the 360-degree camera, which photographs the aurora, is a case of pure knowledge for its own sake.

The geographical South Pole itself stands about 500 yards away from the base. As I walked towards it I could see tiny crystals floating in the air which shone like fireflies in the direction of the sun. There it was: a calculated spot on the snow. An invisible goal. Almost an abstract objective. So little for a man to pin his ambition to.

Amundsen and his men had to check their position with instruments before they knew they'd arrived. As for my father's party, the black dot of Amundsen's tent on the horizon not only told them they had got to the Pole, but also that they'd been forestalled. My father wrote in his diary on that day, 17 January 1912: 'There's a curious damp feeling in the air which chills one to the bone. Great God this is an awful place and terrible enough for us to have laboured to it without the reward of priority.'

And now that I've seen that place I think I know a little better how it must have felt to have to start that 800-mile return journey walking, sledge-hauling across the endless, featureless plateau. From that point my father's diary is really a detailed record of five men who marched on through the blizzards and cold of one of the worst Antarctic summers ever recorded – until they died. Weakened by deficiencies in their diet, they gradually lost the physical strength to pull the heavy sledge, and to withstand the cold. Gradually the rate of advance began to drop: from a dozen miles a day it fell to seven, and then to five.

For the last half of the journey they must have known what impossible odds they faced. Then Edgar Evans fell and hit his head, and not long afterwards he died. Then came the day when Oates walked out into the snow so as not to hold back his companions. And finally a blizzard pinned the last three down in their tent – Birdie Bowers, Bill Wilson and my father. The blizzard swirled endlessly round the tent – and inside their strength ebbed away.

Seventeen years passed before the empty polar plateau was disturbed again, this time by the sound of aircraft engines. In a three-engined Ford monoplane, the greatest of all American polar explorers, Richard E. Byrd, made the first flight to the Pole and back in November 1929. It was done at an average of ninety miles an hour. Byrd wrote afterwards: 'One gets there, and that is about all there is for the telling. It's the effort that counts.'

Another twenty-eight years passed before the Commonwealth Trans-Antarctic Expedition set out to cross the continent by land via the South Pole. But this time with vehicles. Each time the surface collapsed beneath the weight of a snow cat, revealing the yawning depths underneath, precious hours were lost; and for a time, failure seemed to stare them in the face. But once the polar plateau was reached, Vivian Fuchs and his party were able to make up the lost time in style, and on 20 January 1958, Pole Station appeared on the horizon. They were just three days later in the year than my father's party back in 1912. But oh, such a different welcome!

Edmund Hillary was there already, along with a party of

reporters who had flown in to record the great moment. But I doubt if anyone who has ever stood there has quite known the desolation of spirit which my father and his small band of companions must have known on that day in 1912 – or perhaps ever will again.

And now there's a night club at the South Pole: a flourishing game of dice for the regular inhabitants; the inevitable pin-ups decorate the bar. Whatever your reasons for going to the South Pole – even if you've only flown in like a tourist, as I did, you can have a certificate to prove it. I'm now a Knight of the Emperor Penguin Court, by courtesy of the US Navy. Times change.

The Antarctic Treaty itself is a unique achievement. The twelve nations doing active work on the continent have created a truly non-selfish society, where there are no man-made secrets, no military installations, and all territorial claims are held in abeyance. Where else in the world could a group of Americans land at a Russian base and be greeted, first and last, as fellow scientists and human beings? Where else do the Russian and American flags fly from the same pole?

They do at Vostok, the coldest and one of the most isolated places in the world, where the Russians have recorded a temperature of minus 127 degrees Fahrenheit, the lowest ever recorded on earth. On the occasion I was there it was a mere minus 53 degrees. But the warmth of the Russian greeting made up for that. The Americans had come to pick up two of their scientists who had been studying there. It was an occasion for vodka and caviare. This was the Antarctic Treaty in operation. It hasn't had to face a major crisis yet, like the discovery of gold or oil, but the groundwork is there. Antarctica is the only large, truly international territory on earth. Not just an example to the rest of the world, but a blueprint, perhaps, for the control of outer space. For anyone who cares to look around, this Russian base, like all the others in Antarctica, is completely open to inspection.

—

From the top of Observation Hill, behind McMurdo, you can look out due south across the Ross Ice Shelf, which stretches away towards the Beardmore Glacier and the Pole. It was from there that they kept watch, waiting for the polar party that never came. Today, nestling at the foot of that same hill, is McMurdo City, where they think in terms of helicopters and nuclear power and satellite tracking stations – all unknown and undreamed of in my father's day.

And down by the water's edge, that other sure sign of the arrival of man – the garbage heap. The first time the McMurdo dump was set alight the skuas, who had never experienced fire, flew straight into the flames. And even as the fire died down, the survivors alighted on the red-hot cinders. The whole area was covered with dead and dying birds. Now the skuas know better. And perhaps man knows a little better too.

From BBC TV broadcast, 1966

49

Happy the Man

Some time ago I was rash enough to say that I thought I was about the luckiest man I know, and I still think it's true. I was born with an optimistic disposition – a happy disposition – and I've been blessed with good health. I've been fortunate enough to earn my living doing the things I like doing most, being a painter and a naturalist; and most important of all, I've been tremendously happily married for the last fifteen years, perhaps the greatest good luck that a man can have. My life hasn't all run smoothly though. There've been a good many ups and downs, and no doubt there will be more to come.

I was born in 1909, which was the year Blériot made the first cross-Channel flight, and I grew up with a sort of background of the Antarctic. I was only two-and-a-half years old when my father died on his way back from the South Pole, but he wrote letters as his party lay snowbound in their tent, hoping against hope for the blizzard to abate. And these letters were found the following spring. In one of them he wrote to my mother: 'Make the boy interested in natural history if you can.'

Well, children don't always do what their parents want them to, but my mother must have been very clever about it because she succeeded in making me interested in natural history. She had taken a coastguard's cottage at Sandwich and we used to walk along the shore finding shells and whelks' eggs and skates' eggs and sandhoppers and also those glass balls which fishermen use as floats for their nets. And here I first became conscious of birds. There were waders along the shore and skylarks sang above the dunes as they mounted higher and higher into the sky. And it was here too that I saw my first wild geese, brent geese, flying up the shore towards Pegwell Bay.

London was our real home, a little terraced house in Buckingham Palace Road where my mother had her studio. She was a sculptor and had studied under Rodin in Paris. She had a great gift for likeness in her portraits, so many of the famous people of the day came to our house to sit for her; in the afternoon I used to go off to St James's Park to feed the ducks and after tea I was allowed to play the pianola. Unless, of course, Mr Asquith had come to tea and was tired – then not. Mr Asquith was the Prime Minister, and I little knew that his great grandson would one day marry my daughter and be the father of my first grandchildren. But I loved playing that pianola.

Apart from the pianola, however, I never learnt to play any musical instrument, although music has given me some of the most intense enjoyment of my life. And of all composers, I think perhaps Brahms has given me the greatest pleasure, especially his violin concerto.

I went to a preparatory school near Winchester called West Downs, of which my principal memories are of looking for birds' nests and hawk moth caterpillars and singing. It is reliably reported that I was a perfectly poisonous small boy with an angelic soprano voice. Solos in Chapel and Christmas carols – and at school concerts I used to sing 'Polly Wolly Doodle, All The Day'. It became a sort of regular, and I don't know, who got more bored with it, the audience or I.

Oundle School was famous at that time for being one of the few schools at which you could do science. So to Oundle I went. The Headmaster was Kenneth Fisher, father of my dear old friend James Fisher, and he used to take us out to nearby floods to watch thousands of ducks, mostly wigeon, and teal and shovellers and pintails. There, too, we saw wild white-fronted geese and occasionally bean geese, and the wild geese crept a little further into my system. At Christmas time we used to sing great choral works at Oundle, like Bach's B Minor Mass and Handel's *Messiah*. And throughout the autumn term we all used to go round humming some of the best tunes, which made the school quite an inspiring place to be in.

I had always been able to draw a little; in fact, ever since I can remember. And of course there were art lessons at school. We had a delightful Irish art master, who said something to me which I have never forgotten. I used to draw flowers and stuffed birds and things, and one day, looking at one of my drawings, he said:

'That's all right. I know you can draw a stuffed bird, but what I really want to know is whether you can draw a pudding. If you can draw a pudding really well, you can draw anything. Go and draw a pudding.'

Then a Cambridge period and the discovery of wildfowling. Of course, in those days there wasn't so much difference there is today between the hunter's outlook and the naturalist's. And the thing was, that wildfowling took me into the marshes amongst the birds down in the fens of Cambridgeshire and on the coast of the Wash.

Well, like the cavemen, I wanted to paint the quarry on the walls of my cave and so I painted wild geese. My work at Cambridge was natural sciences – zoology, botany, and geology – but I was also painting more and more and I had a small exhibition at one of the book shops in Cambridge, and my first pictures were reproduced in the magazine *Country Life*.

Then suddenly I wondered if I wasn't missing out on art and the humanities and so I decided to change horses in midstream, and to study History of Art and Architecture for my last year at Cambridge. And I think this did give me a broader outlook altogether.

In 1931 I went off to Germany, to Munich, to be an art student at the State Academy. I lived with the family of the

Professor at the animal painting school at the Academy, and I studied in his classes. The conductor of the State Opera House at Munich was Hans Knappertbusch, and I had a friend who was studying with him. One wonderful night, Richard Strauss came to conduct his own composition *Rosenkavelier* which was an altogether memorable performance with Lotte Lehmann. And after it my friend and I were invited to go round to Knappertbusch's flat to play poker, because Strauss was a tremendously keen card player. Well, Strauss won and I lost quite a lot for a penniless art student, but it was worth it to meet such a man.

After the opera I often used to go with Ruli the art student son of the house where I was staying to a bierkeller in the centre of Munich and one night we heard a speaker, not long out of gaol, who was considered an indifferent imitator of Mussolini. That man was Hitler – but he wasn't taken very seriously at that time, and on the way home afterwards, Ruli remarked: 'How could a man with a moustache like Charlie Chaplin expect to be taken seriously?'

But of course, sadly, Ruli was wrong – although at that time the threat of a second World War seemed very slight indeed. *Thé dansants* were all the rage, and I used to go with Ruli's sister, Musch, to the famous 'Four Seasons' hotel in Munich. In fact, we finally won a dancing competition there and I thought it might be amusing to try and write some dance tunes: one that I invented was played by the band at the hotel (it never had any words, but I called it *The Elephant's Cake Walk*).

Back from Germany to the Royal Academy Schools in London, still learning to paint, and at weekends to the Wash for wildfowling. When I'd completed my training as a painter I went to live at Borough Fen Decoy; I painted a lot of pictures there and, of course, they were mostly of birds and marshes. Later I moved into a lighthouse at the mouth of the River Nene on the Wash and my pictures made me a reasonable living so that I could travel on various wild goose chases – in particular, looking for live ones to put in an enclosure which I built round my lighthouse.

I went to Hungary hoping to find the red-breasted goose – an exquisite little black and white goose with a chestnut red breast – and I did in fact see some amongst perhaps 100,000 migrating white-fronted geese on the puszta – the vast grassy plains in the centre of Hungary.

In the early spring this steppeland was covered with beautiful short green turf, and dotted with shallow pools – absolutely ideal for geese. I stayed at a tiny inn, at Csarda, at Nagyhortobágy. There was a gypsy band there which improvised the music and they wrote a little song about this mad Englishman who'd come all the way to Hungary to look for a goose.

Well, I didn't catch any red-breasted geese to bring back for the enclosure at my lighthouse but I did manage to bring back some live snow geese from Canada, from a small hunting club at St Joachim on the edge of the St Lawrence River, quite near Quebec. The club had an enormous picture window, and from inside we had an unparallelled view of the geese which were feeding and resting and fighting and squabbling just outside.

We watched them for an hour and a half until a dog flushed them and they flew off to re-settle farther out on the marsh, looking like snow-flakes, blue against the pink of the setting sun. The snow geese were impressive, but the picture window through which we'd been watching them impressed me even more. I resolved that one day my home should have just such a window, looking out on ducks and geese and swans. And now I have one, at Slimbridge. All these live geese and ducks that I had collected I kept at the lighthouse. This prototype of the Wildfowl Trust grew to be quite an attraction, and people would drive down from London to see the geese.

I was staying with my old sailing friend, John Winter, on the morning in September 1939 when Neville Chamberlain came to the microphone to give his sombre message.

About a year before I'd been enrolled in the Royal Naval

Volunteer Supplementary Reserve, which was a band of yachtsmen, so off I went to Hove to be trained in a huge underground garage which had been taken over by the Navy and called HMS *King Alfred*, a 'stone frigate' that never went to sea. (Although Lord Haw Haw on the German radio once claimed it had been sunk.) A feature of *King Alfred* was the weekly concert, and one of my friends, Dick Addis, wrote brilliant comic songs. His best song described our life in the Wavy Navy, and I used to sing it with new topical words each week.

The destroyers at that time seemed to be the most active and exciting branch of the Navy, and so I was very pleased when I was posted to HMS *Broke* and in due course became her First Lieutenant, in spite of a continuing tendency to seasickness. I spent two uncomfortable years in the *Broke*, fighting the Battle of the Atlantic – or, as we used to call it, the Battle with the Atlantic, because we used to spend a great deal more time fighting the elements than fighting the enemy. Later I had a flotilla of steam gun boats which formed part of the coastal forces.

I emerged from the war with a tremendous feeling, which I think most people shared and which I still retain, that somehow or other we must never ever allow that sort of thing to happen again. My life had been disrupted by the war, as everyone else's had, and after it was over I was left wondering which way to turn. And then someone asked me if I was interested in standing for Parliament in support of Churchill. I agreed to have a go, and soon found myself adopted as Conservative candidate for the new constituency of Wembley North, in the General Election of 1945.

As far as politics go, I was as green as a mallard's head, but I was willing to work hard and make speeches and I was learning fairly quickly how to deal with the rough and tumble of the hustings. Then came the count. I had lost.

It may be that I never had a narrower escape than that! I immediately went back to the things that I'd loved before the war, the birds and painting.

The lesser white-fronted goose at that time had only been recorded once in Britain. Now I had a theory that this wasn't because it didn't come, but because it was overlooked, due to the difficulty of spotting the very slight differences between the lesser and its cousin, the common whitefront (mainly, the bright yellow eyelids of the lesser).

The obvious place to prove this theory was amongst the biggest flocks of whitefronts in Britain, which are on the Severn Estuary. And so one winter's day in 1945 an old friend, Howard Davies, took me to a pillbox, built to ward off the threat of invasion, but on this glorious occasion being used as a vantage point from which to watch the wild geese.

It was on the second morning in the pillbox that Howard Davies managed to pick out a goose which had yellow eyelids, and was clearly a lesser whitefront. This was one of the most spine-tingling moments of my life: my hunch had been borne out, and the thrill was of the same kind as the thrill one gets from listening to great music.

It was while walking back from finding that little rare goose that I suddenly realised this was the place where I should go and live. That was the beginning of the Wildfowl Trust at Slimbridge.

And so our venture began: to discover more about the birds, to promote conservation and save them from oblivion, to teach more people in the world how these creatures could be enjoyed. Parts of Bach's B Minor Mass, which I sang at school, reflect how I feel about wild geese. There's a place in the sanctus chorus where the fugue comes in: *Pleni sunt coeli* – full are the skies of the glory of God. And aren't the great skeins of wild geese a part of the glory of God?

Heard above the cry of the wind, the calls of wild geese can be incredibly sad. The nightingale and the blackcap and the curlew are nature's soloists – but the geese are her chorus, and they're as rousing a sound as sanctus in the B Minor Mass, with each successive skein bringing in the fugue – *Pleni sunt coeli.*

The BBC invited me to be one of the commentators at the wedding of the Queen, then Princess Elizabeth, to Lieutenant Philip Mountbatten, RN, later created Duke of Edinburgh. My place was on the roof of St Margaret's Westminster, overlooking the entrance to Westminster Abbey, and I was, as I always am in broadcasting, extremely nervous.

I'm afraid that my peace of mind wasn't improved when my assistant, who was none other than Rex Alston, trod on the junction box of the headphones, cutting off our direct communication! Quickly we switched on the portable radio, and we were just in time to take our cue from Wynford Vaughan-Thomas.

At our old London house at Leinster Corner, my mother became ill and later she died. That was in 1947. It was a bad year for me all round, and I plunged myself with increasing vigour into building up the Wildfowl Trust.

My studies of wild geese took me one summer to the breeding grounds of Ross's snow goose, in the far north of Canada. This was my first real experience of Arctic exploration. I remember that soon after we set off into the north in a rather played-out Anson aircraft, on what was perhaps a slightly hazardous journey, we had to stop at the little mining town of Yellowknife on the Great Slave Lake. We were held up by the weather and a repair that had to be made to the aircraft.

To this day I can see again the golden brown expanse of the barrens, dappled with snowdrifts and pools of melt water, and teeming with birds. I see the cheerful faces of the Eskimos and I can almost smell the wet caribou hide. I can see the long-tailed ducks with their haunting call which used to live on the lake beside our camp.

I used to have a rather curious vanity that on no account must I follow in my father's footsteps: polar travel was not for me because I must strike out on my own. Recently the BBC invited me to go to Antarctica, to make a film about how people live in the Antarctic today compared with my father's day, and of course I readily accepted! Of the things that happened down there, while we were making the film, I think the most moving part was going into my father's hut at Cape Evans.

Since 1951 I've taken part in quite a few television programmes. It all started with a film I'd taken of ringing pink-footed geese in Iceland. I showed it in the Festival Hall in aid of the Wildfowl Trust, and I wondered whether it would do for television. Desmond Hawkins, who was at that time producing natural history radio programmes, was about to embark on television production, and I made the suggestion to him one day at lunch in Bristol. That was the beginning of the *Look* series.

The series has been running now for more than ten years, with over 150 editions. From Heinz Sielmann's woodpeckers to baby pandas behind the bamboo curtain, from garden birds to 'unknown animals'.

I'm pleased that *Look* has gone so long. Partly because I enjoy making the programmes, but mainly because I believe they've helped people to grasp what we stand to lose if our modern civilisation sweeps all the wildlife away, destroys all the wilderness. A few years back we started the World Wildlife Fund to tackle this problem on an international scale. Living as I do with birds, painting them and studying them, I've developed a great reverence for nature. The extinction of any species at the hands of man strikes me as an appalling thing because it's so final. One of the end products of two thousand million years, twenty million centuries of evolution, is suddenly snuffed out like a candle flame, by a draught which could, nine times out of ten, have been prevented – and in this case there are no matches to rekindle it.

We helped to prevent one such tragedy at Slimbridge – the Hawaiian goose, the ne-ne. By breeding from three original birds, we've now raised over three hundred during the past fifteen years and we've sent ninety back to Hawaii.

Well that's just one small example, but conserving wildlife for the long term benefit of mankind is an enormous task, a majestic task, I think, because nature can be an everlasting source of inspiration to mankind – of spiritual refreshment.

The wild swans which come each year 2,000 miles from Siberia to the little pond at Slimbridge symbolise an ideal relationship between man and nature. They survive the winter more easily because of the food we give them, and certainly they're a part of our family life, our enjoyment of life. And as they sweep down to land, shining, golden under the floodlights in the winter dusk, I realise how greatly they and their kind have contributed to my happiness.

Based on BBC TV broadcast, 1967

Dates and Sources of Original Publication

1. Brown and Green
   From *Morning Flight: A Book of Wildfowl* (Country Life, 1935)

2. A Rare Meal
   From *Wild Geese and Eskimos: A Journal of the Perry River Expedition of 1949* (Country Life, 1951)

3. The Ethics of Shooting
   From *The Eye of the Wind: An Autobiography* (Hodder & Stoughton, 1961)

4. Badgers and the Animals of the Wiltshire Woods
   BBC National Programme broadcast, August 1939

5. The Introduction of Exotic Species
   Address to International Union for Conservation of Nature Technical Meeting, Lucerne, 1966

6. Food Production and Conservation
   Foreword to Elspeth Huxley, *Brave New Victuals: An Inquiry into Modern Food Production* (Chatto & Windus, 1965)

7. Search for Salvadori's Duck
   From *Animals* magazine

8. Skin-Diving: A New World
   From *Animals* magazine

9. Sightseeing in East Africa's Game Parks
   From *Animals* magazine

10. Conservation and Africa
    Introduction to Peter and Philippa Scott, *Animals in Africa* (Cassell, 1962)

11. To Record the Moment – or Enjoy It?
    From *Wild Geese and Eskimos: A Journal of the Perry River Expedition of 1949* (Country Life, 1951)

12. Anabel
    From *Wild Chorus* (Country Life, 1944)

13. The Mysterious Sense of Direction
    From the *London Mystery Magazine* (1949)

14. The Aura
    From *Morning Flight: A Book of Wildfowl* (Country Life, 1935)

15. European Travels
    From *Wild Chorus* (Country Life, 1944)

16. Conscience
    From a letter to the *Geographical Magazine*

17. The Guides Return
    From Peter Scott and James Fisher, *A Thousand Geese* (Collins, 1953)

18. The Flowers of the Eagle's Fell
    From Peter Scott and James Fisher, *A Thousand Geese* (Collins, 1953)

19. Slimbridge Discovered
    From *The Eye of the Wind: An Autobiography* (Hodder & Stoughton, 1961)

20. Bewick's Swans at Slimbridge
    From 17th Annual Report of the Wildfowl Trust, 1964–5

21. How Much Does It Matter?
    From *The Launching of a New Ark: First Report of the President and Trustees of the World Wildlife Fund, 1961–1964*

22. A Voice Crying in the Wilderness?
    From *Happy the Man: Episodes in an Exciting Life* (Sphere, 1967)

23. The Best the Arctic Can Offer
    From *Happy the Man: Episodes in an Exciting Life* (Sphere, 1967)

24. Wild Music
    From *Morning Flight: A Book of Wildfowl* (Country Life, 1935)

25. South America – 1953
    From 6th Annual Report of the Wildfowl Trust, 1952–3

26. Saving a Goose from Extinction
    'To Save the Hawaiian Goose', *The Times*, June 1952

27. Galapagos Expedition
    From 11th Annual Report of the Wildfowl Trust, 1958–9

28. September in the Country
From BBC Home Service broadcast, September 1939

29. Nature of Fear
From *Argosy* (March 1960)

30. Camouflage
From *The Eye of the Wind: An Autobiography* (Hodder & Stoughton, 1961)

31. Fire at Sea
From *The Eye of the Wind: An Autobiography* (Hodder & Stoughton, 1961)

32. A Battle – and a Collision
From *The Eye of the Wind: An Autobiography* (Hodder & Stoughton, 1961)

33. Dawn Off Cherbourg
From *The Eye of the Wind: An Autobiography* (Hodder & Stoughton, 1961)

34. The Story of SGB9
From *The Eye of the Wind: An Autobiography* (Hodder & Stoughton, 1961)

35. 'Gold C' Completed
From *Sailplane and Gliding* (August 1958)

36. Straight and Level, Please
From *Sailplane and Gliding* (October 1958)

37. To Make the Spirit Soar
From a review of Derek Piggott, *Gliding: A Handbook on Soaring Flight*, *Sunday Times*, 1958

38. Prince of Wales's Cup and Olympic Games
From *The Eye of the Wind: An Autobiography* (Hodder & Stoughton, 1961)

39. The Trapeze
From *The Eye of the Wind: An Autobiography* (Hodder & Stoughton, 1961)

40. The America's Cup
Foreword to Robert N. Bavier, Jr, *A View from the Cockpit: Winning the America's Cup* (Dodd, Mead & Co., 1966)

41. The Spirit of Adventure
From *Happy the Man: Episodes in an Exciting Life* (Sphere, 1967)

42. Mr Spriggs and the Crane
From *Cornhill Magazine* (October 1934)

43. Britain's Canals
From Robert Aickman, *Know Your Waterways* (Geoffrey Dibb, 1957)

44. Reflections
From *Happy the Man: Episodes in an Exciting Life* (Sphere, 1967)

45. The Boy David
From *The Eye of the Wind: An Autobiography* (Hodder & Stoughton, 1961)

46. World Champion Skater?
From *The Eye of the Wind: An Autobiography* (Hodder & Stoughton, 1961)

47. Venice, 1939
    From *The Eye of the Wind: An Autobiography* (Hodder & Stoughton, 1961)

48. Journey to the South Pole
    From BBC TV broadcast, 1966

49. Happy the Man
    Based on BBC TV broadcast, 1967

Acknowledgements from the 1967 edition
The publishers would like to thank the following for their kind permission to reprint material which has previously been published or broadcast elsewhere: *Animals* magazine; *Argosy* magazine; the British Broadcasting Corporation; Cassell & Co. Ltd; Chatto & Windus Ltd and Elspeth Huxley; William Collins, Sons & Co. Ltd and James Fisher; William Collins, Sons & Co. Ltd and the World Wildlife Fund; Country Life Ltd; Dodd, Mead & Company Inc.; Geoffrey Dibb Ltd and Robert Aickman; Hodder & Stoughton Ltd; *Sailplane and Gliding* magazine; *The Sunday Times*; *The Times*; and the Wildfowl Trust.